Characterization of SAR Clutter and Its Applications to Land and Ocean Observations

Gui Gao

Characterization of SAR Clutter and Its Applications to Land and Ocean Observations

国防工業出版社
National Defense Industry Press

Springer

Gui Gao
National University of Defense Technology
Changsha, China

ISBN 978-981-13-1019-5 ISBN 978-981-13-1020-1 (eBook)
https://doi.org/10.1007/978-981-13-1020-1

Jointly published with National Defense Industry Press, Beijing, China
The print edition is not for sale in China Mainland. Customers from China Mainland please order the print book from: National Defense Industry Press, Beijing.

Library of Congress Control Number: 2018967735

This Springer imprint is published by the registered company Springer Nature Singapore Pte Ltd.
The registered company address is: 152 Beach Road, #21-01/04 Gateway East, Singapore 189721, Singapore

Preface

Spaceborne and airborne synthetic aperture radar (SAR) sensors have been intensively applied to map, monitor, and analyze the Earth: New operational modes increase the flexibility of SAR sensors that are now able to obtain microwave 2-D images as well as 3-D interferometry products within a wide range of space and time resolution and coverage. However, the SAR data processing is challenging due to the changing properties of targets (such as vehicles, ships, and buildings) and terrains for the electromagnetic waves with various bands, views, polarimetric modes, and configurations. Hence, the definition of new techniques and algorithms for SAR data usage as well as assessment of existing methods for SAR products exploitation are required.

Statistical modeling of SAR images is one of the basic problems of SAR image interpretation. It involves several fields such as pattern recognition, image processing, signal analysis, probability theory, and electromagnetic scattering characteristics analysis of targets. At present, one of the major strategies of SAR image interpretation is to use the methods of classical statistical pattern recognition which are based on Bayesian theory. To utilize these methods for SAR image interpretation, a proper statistical distribution must be adopted to model SAR image data.

The main purpose of this book is to discuss detailed models and applications of statistical models for single-channel and multi-channel SAR images. We summarize and emphasize especially the new developing models based on our work. Meanwhile, some important applications employing these developing models, such as detection and terrain classification, are also reported in this book to assess these statistical models. We expect this book could bring some positive aspects for influencing the forthcoming SAR interpretation. Moreover, this book also provides an index for the researchers, as well as to advance the exploitation of their data for monitoring applications.

We would like to thank Dr. Xianxiang Qin, Dr. Jianghua Cheng, Dr. Gongtao Shi, and Dr. Shujin Sun, for their contributions to the materials included in this book. We appreciate the editors of publication for their efforts and works. This book could not have been completed without their significant contributions. Meanwhile, many thanks must be given to several free Web site for their downloading service of data.

Thanks to my parents, my wife (Mrs. Juan He), and our lovely daughter (Ningqian Gao). Their love is the driving force of all my inspiration and energy.

Changsha, China
December 2018

Dr. Gui Gao

Contents

Chapter 1
Overview for Statistical Modeling of SAR Images

1.1 Introduction

Statistical modeling of SAR images is one of the basic problems of SAR image interpretation. It involves several fields such as pattern recognition, image processing, signal analysis, probability theory, and electromagnetic scattering characteristics analysis of targets etc. [1]. Generally speaking, statistical modeling of SAR images falls into the category of computer modeling and simulation. At present, one of the major strategies of SAR image interpretation is to use the methods of classical statistical pattern recognition which are based on Bayesian Theory and can reach a theoretically optimal solution [1, 2]. To utilize these methods for SAR image interpretation, a proper statistical distribution must be adopted to model SAR image data [1, 2]. Therefore, in the past ten years, statistical modeling of SAR image has become an active research field [1].

Statistical modeling is of great value in SAR image applications. Firstly, it leads to an in-depth comprehension of terrain scattering mechanism. Secondly, it can guide the research of speckle suppression [3–9], edge detection [10], segmentation [1, 11–13], classification [14–16, 17], target detection and recognition [18–22, 23, 24] for SAR images, etc. Finally, combining statistical model with ISAR target database can simulate various SAR images with variable parameters of aspect, terrain content, region position and SCR (signal to clutter ratio), so statistical modeling can provide numerous data for developing robust algorithms of SAR image interpretation [25].

The research on statistical modeling of SAR images may be traced back to the 1970s. With the acquisition of the first SAR image in the U.S., analysis of the real SAR data directly promoted the development of statistical modeling techniques. The speckle model of SAR images, proposed by Arsenault [26] in 1976, is the origin of these techniques, which established the theoretical foundation of the later researches. In 1981, Ward [27] presented the product model of SAR images, which took the speckle model as a special case. As a landmark of the development of statistical modeling, the product model simplified the analysis of modeling. Since then, many

G. Gao, *Characterization of SAR Clutter and Its Applications to Land and Ocean Observations*, https://doi.org/10.1007/978-981-13-1020-1_1

scholars joined this research area and many statistical models of SAR images had been developed.

Since the 1990's, with the coming forth of a series of airborne or space-borne SAR platforms, the acquisition of SAR data is no longer a problem. Due to the urgent demands for analyzing and interpreting the obtained image data, statistical modeling has drawn much attention and become an active research area.

In recent years, many famous research organizations are doing research on SAR statistical modeling [28], and great progress has been made. According to the collected literatures, from 1986 to 2004, there are more than 100 papers dealing with SAR statistical modeling published in some famous journals such as *IEEE-AES, IEEE-IP, IEEE-GRS,* and *IEE*, etc. and in some international conferences such as SPIE and IGARSS. The related papers, which use SAR statistical model for the purpose of segmentation, speckle suppression, classification and target detection and recognition, are uncountable. Much creative research has been made. Proc. Oliver, an English scholar, published his monograph *Understanding Synthetic Aperture Radar Images* in 1998 [1]. The book includes 14 chapters, two of which discuss the statistical modeling technology. It summarizes related techniques on SAR statistical modeling before 1997. After 1997, literatures on SAR statistical modeling can be seen in the famous journals almost every year. The most attractive achievement among them is the statistical modeling on extremely heterogeneous region of SAR images proposed by Frery [28] who works in Brazil. the originate ideas are introduced that SAR image can be divided into homogeneous regions, heterogeneous regions and extremely heterogeneous regions according to their contents for the purpose of statistical modeling. Further more, statistical modeling of SAR images is taken as one of the main contents in more than 20 doctors' dissertations of UMI and in the research reports from Belgium Royal Military Academe. While numerous statistical distributions have been proposed to model SAR image data, we are unaware of any surveys on this particular topic. It is necessary to categorize and evaluate these models and relevant issues.

The main contribution of this survey is the classification and evaluation of the statistical models of SAR images existed currently. The vital and latest contributions have been covered also. The survey is organized as follows. Section 1.2 illustrates the classification and the research contents of statistical modeling. In Sect. 1.3, current statistical models are discussed in detail. The relationship of them and their limitations in applications are pointed out. Major conclusions and developing trends of statistical modeling are also presented. We conclude the survey in the final section.

1.2 Model Classification and Research Contents

According to the modeling process, the statistical models of SAR images can be divided into two categories [2, 29, 30, 31, 32]: parametric models and nonparametric models. When dealing with a parametric model, several known probability distributions of SAR imagery are given at first. Usually, the parameters of these distributions are unknown and have to be estimated according to the real image data. Finally, by using some certain metrics, the optimal distribution is chosen as the statistical model of the image. While handling a nonparametric model, no distributions have to be assumed, and the optimal distribution is obtained in a way of data-driven of image data. The merit of the nonparametric models is that they make the process of statistical modeling more flexible and can fit the real data more accurate.

Since nonparametric modeling involves complex computation as well as numerous data, it is usually time-consuming and cannot satisfy the requirements of various applications [29]. Consequently, parametric modeling is intensively studied. The process of parametric modeling can be described in brief as to choose an appropriate one from several given statistical distributions for the image to be modeled. The process is shown in Fig. 1.1. According to Fig. 1.1, the process of parametric modeling consists of: (1) analyzing several known statistical distribution models; (2) parameter estimation: estimating the parameters of different distribution; (3) goodness-of-fit tests: assessing the accuracy of the given models fitting to the real data.

1.2.1 Parameter Estimation

Several strategies have been proposed in the literature to deal with parameter estimation [30]. The two most frequently used methods are probably the "method of moments" (MoM) [1, 17, 33] and the maximum-likelihood (ML) methodology [23, 31, 34]. Recently, the method of log-cumulants (MoLC) is also included as a possible parameter estimation approach [3, 17, 35].

1.2.2 Goodness-of-Fit Tests

A number of methods for quantitatively assessing the validity of statistical models in light of sample data have been developed over the last hundred years. Many of these

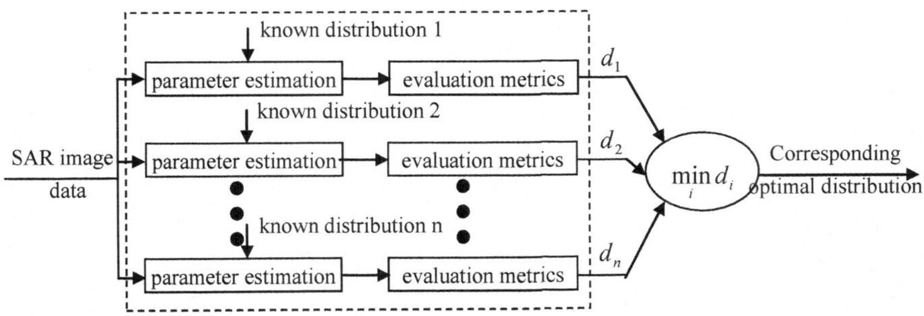

Fig. 1.1 A general flow chart of parametric modeling

methods place the problem in a statistical hypothesis testing framework, pitting a null-hypothesis H_0, an assertion that the data were not generated according to the model, against an alternative hypothesis H_1, an assertion that they are not. The methods are then implemented by computing some statistic of the random observations that has a known distribution if H_0 were true. Values of this quantity close to zero are interpreted as evidence that H_0 should be rejected in favor of H_1. The purpose of these methods is to seek the model that best describes observed data from a set of specified models, irrespective of whether any model is actually a good fit to the data [36].

In summary, the major rules for assessing the fitting accuracy includes χ^2 matching test [36, 37], AIC (Akaike information criteria) rule [38], *K–S* (Kolmogorov–Smirnov) test [36, 39, 40], *K–L* distance measurement [41, 42], D'Agostino-Pearson test [2, 36, 43], and Kuiper tests [35] etc. The research on parameter estimation as well as accuracy assessment is relatively mature and will not be discussed further here. Associated literatures [2, 36, 35] can be read for more information.

1.3 Statistical Models

The purpose of statistical modeling of SAR images is to determine a statistical model for single-polarimetric images or multi-polarimetric images. The multi-polarimetric SAR images are a combination of four basic kinds of polarimetric images represented by the scattering matrix. For any one of the polarimetric image, its statistical characteristic has no difference from a single-polarimetric image. The single-polarimetric statistical model can be extended to describe the multi-polarimetric images [44–47]. Therefore, studying the statistical models of single-polarimetric SAR images is of basic significance. This section mainly discusses this kind of models.

It is more than 30 years since the SAR statistical model has been first studied. Researchers have proposed various statistical models, among which the statistical model family based on the product model outperforms other models [2]. So we would like to comprehensively summarize current statistical models using the product-model-based ones as a thread.

1.3.1 Nonparametric Models

The nonparametric models are an effective kind of models which can estimate the probability density function (PDF) of SAR image data based on the nonparametric method. The basic idea is to use the weighted sum of different kernel functions to obtain the estimation of the statistical distribution. Typical methods include: the Parzen window technique [31, 48, 49] the artificial neural networks (ANN) method [50, 51], the support vector machine (SVM) method [52, 53, 54] etc. The characteristic of the nonparametric models is that it is a data-driven model and suitable for estimating the complex unknown PDF. Nonparametric modeling has high estimation accuracy, but it usually needs a large sample data set as well as complex operations and is a time-consuming task. Consequently, it's seldom used in any applications, except several reports focus on the problem of ship target detection in SAR images with simple sea backgrounds [48].

1.3.2 Parametric Models

The underlying idea of parametric modeling is to use the parameter estimation method to determine the statistical model of SAR image data according to some known distributions. During the past 20 years, the parametric model has been widely and thoroughly studied. With the analysis of data from different sensors and the scattering mechanism of different kinds of terrain, many concrete SAR statistical distributions for different cases have been proposed.

1.4 Classification of Parametric Models

The parametric models can be classified into four categories according to its main idea (see Fig. 1.2): (1) the empirical distributions; (2) the models developed from the product model (PM); (3) the models developed from the generalized central limit theorem (GCLT); (4) other models.

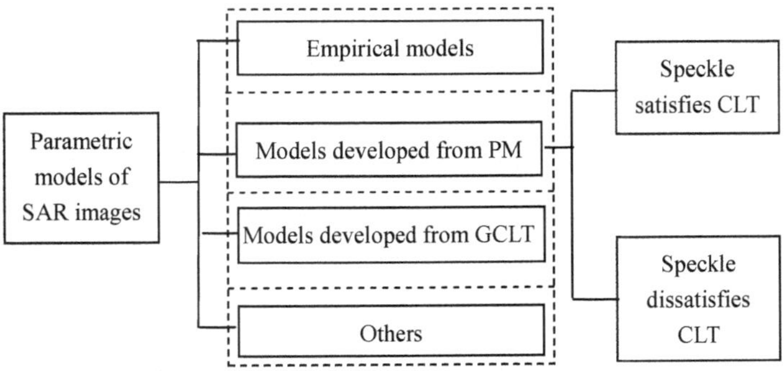

Fig. 1.2 Four major categories of parametric modeling. Note: PM represents the product model; CLT represents the central limit theorem; GCLT represents the general central limit theorem

1.4.1 The Statistical Models Developed from the Product Model

The product model is widely used in SAR image analyzing, processing and modeling. Most of the widely-used statistical models are developed from the product model, which derived from the speckle model. The process of developing concrete statistical models from the speckle model is shown in Fig. 1.3.

The speckle model, coming from the random walking model (see Fig. 1.4) and proposed by Arsenault [26], is deduced from the coherent imaging mechanism of a SAR system, under the ideal circumstance that the imaged scene has a constant RCS (i.e. speckle is fully developed and homogeneous surfaces appear as stationary fields). The deducing process based on the coherent imaging mechanism begins with the six reasonable hypotheses as follows [1, 30, 55, 56]:

- Each resolution cell contains sufficient scatterers;
- The echoes of these scatterers are independently identically distributed;
- The amplitude and phase of the echo of each scatterer are statistically independent random variables;
- The phase of the echo of each scatterer is uniformly distributed in $[0, 2\pi]$;
- Inside a resolution cell, there are no dominant scatterers;
- The size of a resolution cell is large enough, compared with the size of a scatterer.

Secondly, with the six hypotheses mentioned above and the central limit theorem (CLT) [57], it can be proved that the energy of each resolution cell has a negative

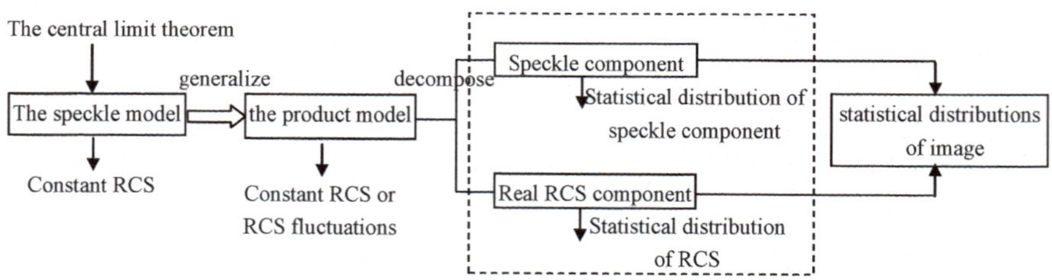

Fig. 1.3 Process of developing a statistical model from the product model

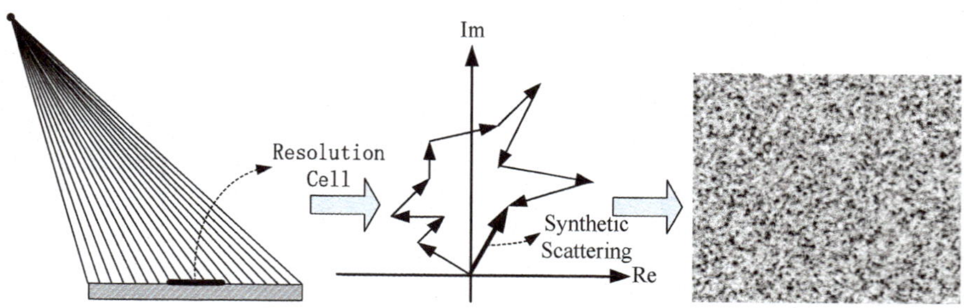

Fig. 1.4 Random walking model

exponential distribution with the mean value equal to the real RCS value of the resolution cell. Finally, according to the hypothesis of constant RCS background, each resolution cell can be considered as a stochastic process, with the ergodic property (i.e., each resolution cell is statistically independent). Therefore, the whole image has a distribution identical to that of a single resolution cell.

Motivated by the speckle model, Ward [27] proposed the product model of SAR images. Figure 1.3 shows the process of developing a statistical model from the product model. According to Fig. 1.3, the product model combines an underlying RCS component σ with an uncorrelated multiplicative speckle component n, so the observed intensity I in a SAR image can be expressed as the product [42, 58–62]:

$$I = \sigma \cdot n \tag{1.1}$$

The speckle model is taken as the special example of the product model with a constant RCS (σ). Because the product model is correlated with the underlying terrain RCS (σ), it is usually applied to the intensity data (energy or the square of amplitude). That is, I in Eq. (1.1) represents the observed value of the intensity. The product model simplifies the analysis of the statistical model of SAR images. So it is widely used to develop models which take the RCS fluctuations into consideration.

Since the speckle component has a determinate statistical distribution, only the RCS fluctuation component need to be considered when developing the statistical models of SAR images (see Fig. 1.3). According to the product model in Eq. (1.1), the PDF of the observed intensity is given by

$$P(I) = \int_{0}^{\infty} P(\sigma)P(I|\sigma)d\sigma \tag{1.2}$$

where $P(\sigma)$ represents the RCS component distribution and $P(I|\sigma)$ is correlated with the distribution of speckle component.

Figure 1.5 gives the statistical models of constant RCS or RCS fluctuations when the speckle component satisfy the central limit theorem. As Fig. 1.5 shown, many classical statistical models, called the family of Gaussian model, have been derived based on the speckle model, a special example of the product model. Either in the high-resolution or low-resolution case, with the hypothesis of a constant RCS background and the central limit theorem, both the I and Q components of the speckle are Gaussian distributed with unit mean. Thus, as is shown in Fig. 1.5, the single-look amplitude has a Rayleigh [1] distribution; the single-look intensity has a negative exponential distribution [1] with unit mean; the multi-look amplitude has a square root Gamma distribution; the multi-look intensity has a Gamma (or Nakagami-Gamma) [1, 30, 32, 63] distribution with unit mean, etc.

The RCS of a homogeneous region (e.g. the grassland region) in either low-resolution or high-resolution SAR images can be expected to correspond to a constant. Actually, most scenes contain in-homogeneous regions with RCS fluctuations [1, 30, 55]. According to Jakeman and Pusey's [64] investigation, when the number of

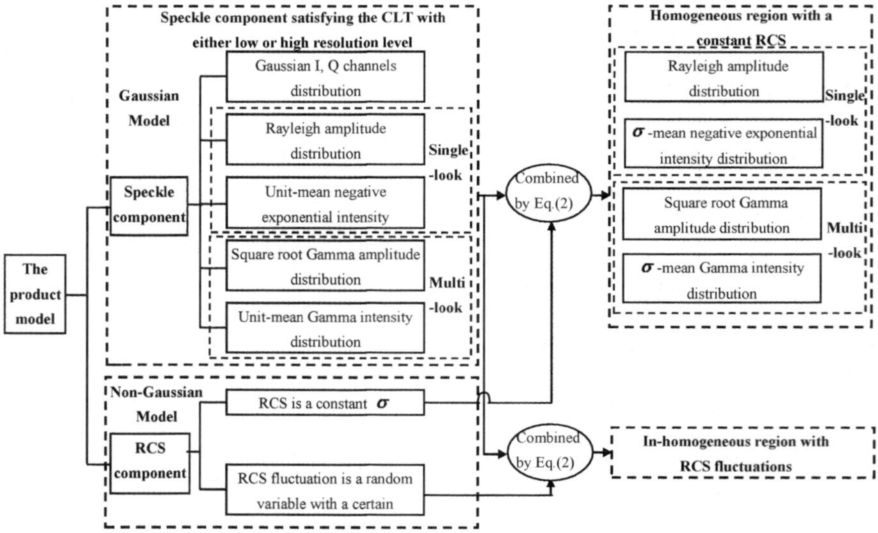

Fig. 1.5 Statistical models of constant RCS or RCS fluctuations when the speckle component satisfy the central limit theorem

scatterers in a resolution cell becomes a random variable due to fading phenomenon and the population of scatterers to be controlled by a birth-death-migration process, it should have a Poisson distribution [1] and the mean of the Poisson distribution in each resolution cell (i.e. the expected number of scatterers) itself is also a random variable [28, 40, 65, 66]. If the mean is Gamma distributed, the corresponding intensity data should have a *K* [1, 34, 64, 67–71] distribution. A further research indicates that *K* distribution can be viewed as the combination of two split parts according to Eq. (1.2) in the framework of the product model [1]: (1) the speckle component satisfying the central limit theorem; (2) the Gamma distributed intensity RCS fluctuations. The *K* distribution is deduced with the assumption that the underlying intensity RCS fluctuations have a Gamma distribution in a heterogeneous region. The Gamma distribution can well describe the characteristics of the RCS fluctuations of a heterogeneous terrain in high-resolution SAR images. The deduced *K* distribution itself has the multiplicative fading statistical characteristics and usually provides a good fit to the heterogeneous terrain. Therefore, the *K* distribution has become one of the most widely used and the most famous statistical models in recent years [64, 72, 73]. Some extensive applications of *K* distribution can be found [40]. Oliver proposed correlated *K* distribution [65]; Jao used *K* distribution in the case of rural illuminated [72]; Barakat obtained *K* distribution in case of weak scattering [74]; and Yueh created and extension of *K* distribution for multipolarization images [66]. Furthermore, according to the deducing process of the *K* distribution, the homogeneous region with a constant RCS can also be described as a special case of the *K* distribution [1]. MoM turns out to be feasible for the parameter estimation task concerning a K-distributed random variable [68, 69], whereas no closed form is currently available for ML parameter estimation [34, 69], thus requiring intensive numerical computations or analytical approximations of the PDF itself [1, 30].

Motivated by the derivation of K distribution, Delignon [40, 75] proposed that when the expected number of scatterers in every resolution cell has an inverse Gamma intensity distribution [40, 75], a Beta intensity distribution of the first kind [40, 67, 75] or a Beta intensity distribution of the second kind [40, 67, 75], the corresponding heterogeneous region will has a B, U or W distribution respectively (i.e., the Pearson system of parametric families [17, 75). Similarly, these three kinds of intensity distribution models can be seen as the combination of the speckle component and the terrain RCS intensity component in the framework of the product model expressed as Eq. (1.2). Figures 1.5 and 1.6 show the statistical models when the speckle component satisfies the central limit theorem.

The K, U and W distributions have been reported to be appropriate for the heterogeneous terrain such as the woodland and the cultivated cropland. But they cannot meet the demand for the statistical modeling of complex scenes in high-resolution images. The complexity of the high-resolution scenes mainly lies in two aspects [55]: (1) the terrain of the scene is usually extremely heterogeneous, such as the urban region containing many buildings, which results in the severe long-tailed part of the image histogram; (2) there exist two or more heterogeneous components in a certain scene, such as a combination of woodlands and grasslands, etc.

To solve these problems, Frery deduced a new statistical model, the G model [23, 28, 76–79] based on the product model assuming a Gamma distribution for the speckle component of multi-look SAR images and a generalized inverse Gaussian (GIG) law for the signal component [28, 30, 78, 80], as is shown in Fig. 1.6. It is Frery

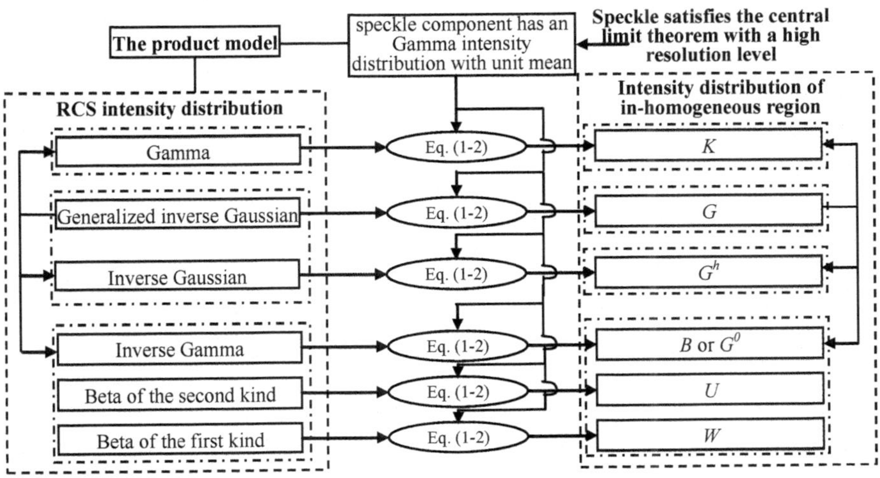

Fig. 1.6 Statistical models of RCS fluctuations when the speckle component satisfies the central limit theorem. Note: 'A⇒B' means 'B is a special example of A'

who first proposes to divide a region as homogeneous, commonly heterogeneous or extremely heterogeneous according to its homogeneous degree when deducing the G model. The K and G^0 (also called B distribution) distributions are two special forms of the G model. The former is appropriate for the heterogeneous region and the latter is appropriate for the extremely heterogeneous region. The G^0 distribution can be converted into the β' (Beta-Prime) distribution under the single-look condition. Although the G^0 distribution is a specific example of the G model, it has a more compact form in comparison with the G model and consequently has a simple parameter estimation method. The relationship between the G^0 distribution and the K distribution cannot be deduced theoretically. The parameters of the G^0 distribution are sensitive to the homogeneous degree of a region, which makes the G^0 model appropriate for modeling either heterogeneous or extremely heterogeneous region. Moreover, MoM can be easily and successfully applied to parameter estimation of the G^0 distribution. Frery [28, 76] and Muller [77, 78] carried out experiments on many SAR images of different kinds of terrain with various band, polarization, resolution and look numbers, such as different urban areas, homogeneous and heterogeneous regions, etc. Their results testified the good characteristics of the G^0 distribution.

A further particular case of the G model (named the "harmonic branch" G^h assuming that the intensity RCS fluctuations of the background are the inverse Gaussian (IG) distribution which has also been employed to model the intensity statistics [28, 78]) is proposed in [78] and endowed with a moment-based estimation approach to images of urban areas and mixed terrain.

Eltoft [81–83] assumed a normal IG distribution for the real and imaginary parts of the backscattered complex signal, thus resulting in an amplitude PDF (i.e. "Rician inverse Gaussian", RiIG) formulated as a combination of an IG PDF and a Rice PDF(see Section D). The purpose of their investigation is to describe the statistics of ultrasound images. While given the similarities between SAR and ultrasound, RiIG can also be used as a model for SAR images. Finite applications of statistical modeling for SAR images can be found in [83]. Anyway, further experimental investigation using real SAR data is needed.

The above models developed from the product model are all derived under the hypothesis that the speckle component satisfies the central limit theorem. Theoretically, when the resolution becomes high enough, the resolution cell will be so small that the central limit theorem cannot be applied any more. Thus, the above models are not appropriate for modeling of the high-resolution SAR images. Accordingly, Anastassopoulos [37, 84–86] proposed a generalized compound probability distribution (GC distribution, see Fig. 1.7) in which the speckle and intensity RCS fluctuation components theoretically are generalized Gamma distributed (GΓ distribution) [37]. The GC distribution has no analytic expression only with a given integral form, so it is difficult to utilize. With a large number of experiments, we [42] have proven that even if the resolution is high up to 0.3 m, the speckle component still satisfy the central limit theorem. So it is not necessary to adopt the GC distribution for SAR images with a resolution lower than 0.3 m. Besides, due to the absence of the higher-resolution data, further experiments are needed for validating the rationality of the GC distribution.

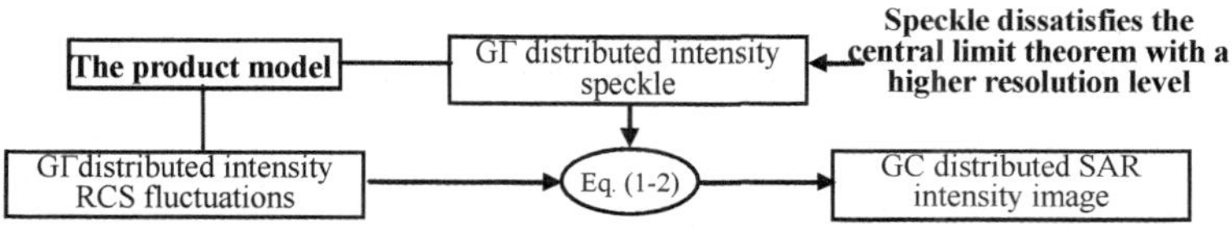

Fig. 1.7 Statistical models when the speckle component dissatisfies the central limit theorem

1.4.2 The Statistical Model Developed from the Generalized Central Limit Theorem

Another thread of statistical modeling is to develop the models based on the generalized central limit theorem [55]. According to the knowledge of probability theory, the generalized central limit theorem states that the sum of a set of independently identically distributed random variables, no matter their variances are finite or infinite, will converge to the α-stable distribution [2, 87–89], which is essentially a more general distribution model. Tsakalides et. al [87] and Pierce [88] therefore considered that the symmetric α-stable distribution (SαS) [90, 91] should be applied to model the real and imaginary parts of the data separately received by the SAR system. The empirical fitting results obtained by Kappor [89] and Banerjee [92] indicated that the SαS distribution could describe some woodland regions in the UWB-SAR images.

In order to consider further the statistical modeling problem of narrowband SAR images, Kuruoglu [3, 55, 93] introduced the generalized heavy-tailed Rayleigh amplitude distribution based on the SαS (here after simply denoted by SαSGR), which can fit the urban SAR images with a long tail. It can be proved that this distribution is a compound Rayleigh distribution [93, 94] and a spherical invariant random process (SIRP) [95]. Although the SαSGR is a more accurate statistical model of SAR images in theory, without any analytic expression. A moment-based estimation strategy is developed in [55] for this parametric model. However, it is very difficult to apply.

1.4.3 The Empirical Distributions

The empirical distributions have no sound deduction in theory. They come from the experience of analyzing real data. Several empirical models have been used to characterize the statistics of SAR amplitude or intensity data, such as Weibull, log-normal, and Fisher PDFs.

The log-normal distribution was proposed by George [96]. Its major motivation was to adopt a homomorphic filter to convert the multiplicative noise in a SAR image to the additive Gaussian white noise with the assumption that the logarithmic SAR image was Gaussian distributed. The log-normal distribution, with a broad dynamic range, is a familiar statistical model which can describe the non-Rayleigh data. But it is a poor representation of the lower part of the SAR image histogram, with the data over-fitted phenomenon [55, 97]. Fukunaga [98] stated that it was inappropriate to fit the logarithmic SAR image to a Gaussian distribution, and that the quarter power domain of the logarithmic data was more consistent with a Gaussian distribution.

The Weibull distribution [99] is also a good statistical model of the non-Rayleigh data. Compared with the log-normal distribution, it can fit the experimental data in a broader range. The Rayleigh distribution and the negative exponential distribution are two special examples of Weibull distribution with specific parameters. Therefore, Weibull distribution can describe single-look images precisely for either amplitude or intensity. Experiences have shown the Weibull distribution cannot represent multi-look images exactly [1].

Recently, the Fisher distribution has also been adopted as an empirical model for the SAR statistics over high resolution urban regions [17]. The Fisher distribution also is proved to be equivalent to a G^0 PDF [17, 30].

1.4.4 Other Models

Goodman [17, 30, 63, 100] presented that when a resolution cell is dominated by a single scatterer, the corresponding intensity image has a Rician distribution (or Nakagami-Rice distribution [1]). Theoretically, in the case of low resolution, when the strong scatterers representing the targets are embedded into the surrounding weak clutter environment, the Rician model is appropriate to describe the corresponding image [63, 101].

Blake [41, 102] introduced a joint distribution model when considering two or more than two heterogeneous terrain types in the scene of a SAR image. Firstly, the optimal statistical model of a homogeneous region is analyzed and the K distribution is proven as the best model by the experiments. Secondly, according to the ratio of each terrain to the whole scene, several K distributions weighted with the ratios respectively are summed up to describe the image. The unknown parameters of the joint distribution model increase several times in number and thus makes the parameter estimation more difficult. Generally, such parameter estimation is based on solving a set of nonlinear equations [36, 68, 103], which will impede the application of the joint distribution.

Blacknell [104, 105] proposed a statistical distribution model considering the correlation between pixels. Since the pixels of a real SAR image are usually dependent, there is certain correlation between the pixels. Blacknell adopted the mixed Gaussian distribution to model the correlation between the pixels and deduced a statistical model. In fact, the mixed Gaussian distribution can describe only the simplest case of the correlation between the pixels. Further researches are expected for more complicated cases [65, 104, 105].

Besides, literatures [106–108] proposed some other models, which are mostly the generalization or modification of the models mentioned above. Limited to the length of the survey, no further discussion is given.

1.5 The Relationship Among the Major Models and Their Applications

1.5.1 The Relationship Among the Parametric Statistical Models

The statistical model of a single-look image is a special example of the corresponding multi-look model when the look number $n = 1$. Let $P_I(I)$ be the PDF of the intensity I and $P_A(A)$ be the PDF of the amplitude, then the following relationship holds [1]:

$$P_A(A) = 2A \cdot P_I\left(A^2\right) \tag{1.3}$$

or

$$P_I(I) = P_A\left(\sqrt{I}\right) \Big/ 2\sqrt{I} \tag{1.4}$$

Hence, the statistical distribution of single-look data can be deduced from that of multi-look data; and the distribution of the amplitude can be deduced from that of the intensity. Additionally, the log-transformed distributions are also deduced easily according to [61]. Based on this conclusion, Fig. 1.8 illustrates the relationship among the current major statistical models. Some other models are not shown in Fig. 1.8 because no theoretical relationship for them can be established to the models in Fig. 1.8. The concrete expressions of various distributions can be seen in [2, 23].

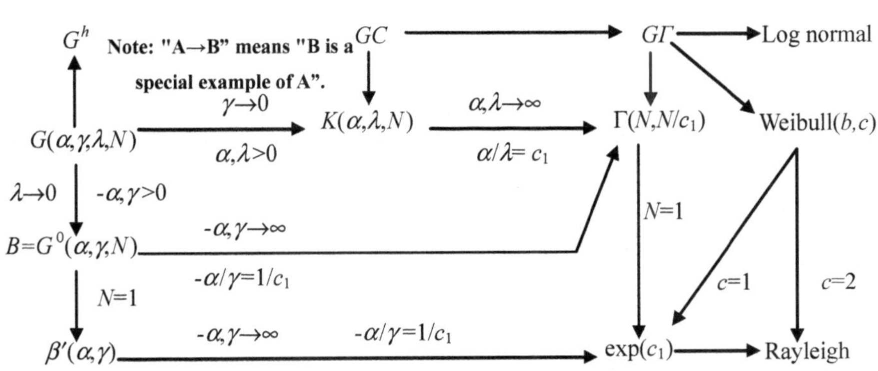

Fig. 1.8 Relationship among the major statistical models (N is the look number)

1.5.2 *Summary of the Applications of the Major Models*

According to many researchers' experiences [1] and the authors' analysis, Table 1.1 summarizes the characteristics and the application areas of the major models discussed in the previous sections.

1.6 Discussion of Future Work

Much progress has been made with the research of statistical modeling of SAR images in the past few tens of years, especially during recent years. The related literatures are uncountable. As far as we could comprehend, the major conclusions and several promising directions for further research are summarized as follows:

(1) Regarding the deducing process of current statistical models, many assumptions are made to acquire the models, so these models can only approximately describe the electromagnetic scattering characteristics of the scene in theory, which is the common shortcoming of all the statistical modeling of the scene. How to construct models that can exactly describe the electromagnetic scattering characteristics of a scene will be a big challenge.
(2) Among the existing statistical models, those developed from the product model are the most widely used and the most promising. This can also be seen from the related literatures.
(3) The statistical models based on the product model can be divided into two cases according to whether the speckle component satisfies the central limit theorem or not. Correspondingly, there are two typical models, i.e. the widely used G^0 model and the GC model with difficulty in application. The problem is, what level on earth the resolution is increased to that the speckle component don't satisfy the central limit theorem any longer. No conclusion has been made yet.
(4) It is a novel idea to model a region according to its homogeneousness degree. The G^0 model (the β' model at single-look case) is the optimal one among the models developed from the product model. On one hand, the parameters of the G^0 model are sensitive to the homogeneousness degree of the observed images. Such a characteristic make it suitable for modeling the homogeneous, heterogeneous or extremely heterogeneous, single-look or multi-look, intensity or amplitude data. That means it can be universally used. On the other hand, many widely used models can be unified to the G^0 model (see Fig. 1.8).
(5) All the statistical models, even the G^0 model, can describe the regions only with relatively simple contents and a few terrain types. In other words, the statistical model has the so-called "regional" characteristic. For the large- scale scene, whose contents are complex and terrain types are extremely numerous, it is impractical to use the statistical models with a few parameters to describe the whole image. However, models with too many parameters also cause difficulties in applications. Therefore, it is a trend to build a statistical model with the "regional" characteristic. Typically, Billingsley [39] assess the fit of Rayleigh,

Table 1.1 Summary of the applications of the major models

Model families	Model	Analytic expression?	Parameter estimation	Application cases	Notes
1	*Weibull*	Yes	Complex	High-resolution, amplitude or intensity, single-look	Unsuitable for multi-look images
	Lognormal	Yes	Simple	Moderately high-resolution, amplitude	Data over fitted phenomenon
	Fisher	Yes	Simple	Homogenous, heterogeneous or extremely heterogeneous region, multi- or single-look, intensity or amplitude	Be equivalent to a G^0 distribution
2	*Rayleigh*	Yes	Simple	Homogenous region, single-look, amplitude	Widely used in interpretation algorithms
	Exp	Yes	Simple	Homogenous region, single-look, intensity	Widely used in interpretation algorithms
	Gamma	Yes	Simple	Homogenous region, multi-look, intensity	The amplitude distribution corresponding to the square root Gamma
	K	Yes	Complex	Moderately heterogeneous region, multi- or single-look, intensity or amplitude (having corresponding expressions for each case)	Widely used in interpretation algorithms
	U, W	Yes	Complex	Moderately heterogeneous region, multi- or single-look, intensity or amplitude (having corresponding expressions for each case)	Seldom used in interpretation algorithms

(continued)

Table 1.1 (continued)

Model families	Model	Analytic expression?	Parameter estimation	Application cases	Notes
	G	Yes	Complex	Homogenous, heterogeneous or extremely heterogeneous region, multi- or single-look, intensity or amplitude (having corresponding expressions for each case)	Difficult to apply
	G^0	Yes	Simple	Homogenous, heterogeneous or extremely heterogeneous region, multi- or single-look, intensity or amplitude (having corresponding expressions for each case)	A special example of the G distribution, also called the *B* distribution, widely used
	β'	Yes	Simple	Homogenous, heterogeneous or extremely heterogeneous region, single-look, intensity	A special example of the G^0 distribution, widely used
	G^h	Yes	Simple	Extremely heterogeneous urban areas and mixed terrian	A special example of the G distribution
	RiIG	Yes	Simple	Ultrasound images	Further investigation for SAR images is needed
	GC	No	Complex	Various image data with an extremely high resolution level	A general form of many other models, difficult to apply, further validation is needed

(continued)

Table 1.1 (continued)

Model families	Model	Analytic expression?	Parameter estimation	Application cases	Notes
3	*SαS*	No	Complex	Real and imaginary components of SAR data	Used in modeling the woodland regions in UWB SAR data
	SαSGR	No	Complex	Long-tailed amplitude image of urban area	Difficult to apply
4	*Rician*	Yes	Complex	Low-resolution image with targets in weak clutter	Seldom used
	Jointly distribution	Yes	Complex	Heterogeneous	Difficult to apply
	Mixed Gaussian	Yes	Simple	Considering the correlation between pixels	Correlation is simple, further research is needed

Note "1" Represents the empirical distributions
"2" Represents the models developed from the product model
"3" Represents the models developed from the general central limit theorem
"4" Represents other models

Weibull, log-normal, and K-distributions to pixel magnitudes in clutter data and show via the K–S test that none fit well over the entire range of magnitudes.

(6) According to the related literatures, once a model was proposed, it would be applied to diverse images with several bands and different view angles. Usually, their results were good. Generally speaking, the diversity of the band and the view angle of a sensor within a certain scope has slight influence on statistical modeling of the SAR data.

(7) It is also a new idea to consider the correlation among the SAR data. In theory, it can expose the statistical characteristics of SAR images more accurately. However, it's hard to exactly model the correlation. Borghys [103] analyzed the effect on the statistical model caused by the correlation among pixels. His conclusion was that through appropriate down sampling, such effect could be ignored when modeling SAR images.

1.7 Conclusions

Statistical modeling of SAR images is one of the basic research topics of SAR image interpretation. It is of great significance both in theory and in applications. Based on an extensive investigation on the related literatures, we begins with the history and current research state of statistical modeling of SAR images. Then, statistical modeling techniques are thoroughly reviewed using the product model as a thread and some major problems are briefly illustrated in order to attract more attentions in this field. We believe that the research will progress widely and deeply due to the demands of SAR image interpretation.

References

1. C.J. Oliver, *Understanding Synthetic Aperture Radar Images* (Artech House, Boston/London, 1998), Chaps. 4, 5
2. Q. Zhang, *Research on Detection Methods of Vehicle Targets from SAR Images based on Statistical Model (in Chinese)*. Master's Dissertation, National University of Defence Technology, China, 2005, Chap. 2
3. A. Achim, E.E. Kuruoglu, J. Zerubia, SAR image filtering based on the heavy-tailed Rayleigh model. IEE Trans. Image Process. **15**(9), 2686–2693 (2006)
4. A. Achim, P. Tsakalides, A. Bezerianos, SAR image denoising via Bayesian wavelet shrinkage based on heavy-tailed modeling. IEEE Trans. Geosci. Remote Sens. **41**(8), 1773–1784 (2003)
5. M. Walessa, M. Datcu, Model-based despeckling and information extraction from SAR images. IEEE Trans. Geosci. Remote Sens. **38**(9), 2258–2269 (2000)
6. R. Touzi, A review of speckle filtering in the context of estimation theory. IEEE Trans. Geosci. Remote Sens. **40**(11), 2392–2404 (2002)
7. F. Zhang, Y.M. Yoo, L.M. Koh, Y.M. Kim, Nonlinear diffusion in Laplacian pyramid domain for ultrasonic speckle reduction. IEEE Trans. Med. Imaging **26**(2), 200–211 (2007)
8. J.S. Lee, Speckle analysis and smoothing of synthetic aperture radar images. Comput. Graph. Image Process. **17**, 24–32 (1981)

9. V.S. Frost, J.A. Stiles, K.S. Shanmugan, J.C. Holtzman, A model for radar images and its application to adaptive digital filtering for multiplicative noise. IEEE Trans. Pattern Anal. Mach. Intell. **PAMI-4**(2), 157–162 (1982)
10. R. Touzi, A. Lopes, P. Bousquet, A statistical and geometrical edge detector for SAR images. IEEE Trans. Geosci. Remote Sensing **26**, 764–773 (1988)
11. P.T. Macri, C.J. Oliver, P. Lombardo, Segmentation-based joint classification of SAR and optical images. IEE Proc. Radar Sonar Navig. **149**(6), 281–296 (2002)
12. J. Lee, I. Jurkevich, Segmentation of SAR images. IEEE Trans. Geosci. Remote Sens. **27**, 674–680 (1989)
13. R. FjØrtØft, Y. Delignon, W. Pieczynski, M. Sigelle, F. Tupin, Unsupervised classification of radar images using hidden Markov chains and hidden random fields. IEEE Trans. Geosci. Remote Sens. **41**, 675–686 (2003)
14. Y. Koukoulas, F.T. Ulaby, L.E. Pierce, The Bayesian hierarchical classifier (BHC) and its application to short vegetation using multifrequency polarimetric SAR. IEEE Trans. Geosci. Remote Sens. **42**(2), 469–477 (2004)
15. A.N. Nyonngui, E. Tonye, A. Akono, Evaluation of speckle filtering and texture analysis methods for land cover classification from SAR images. Int. J. Remote Sens. **23**(9), 1895–1925 (2002)
16. H. Deng, D.A. Clausi, Unsupervised segmentation of synthetic aperture radar sea ice imagery using a novel Markov random field model. IEEE Trans. Geosci. Remote Sens. **43**(3), 528–538 (2005)
17. C. Tison, J.M. Nicolas, F. Tupin, H. Maitre, A new statistical model for Markovian classification of urban areas in high-resolution SAR images. IEEE Trans. Geosci. Remote Sens. **42**(10), 2046–2057 (2004)
18. M.D. Bisceglie, C. Galdi, CFAR detection of extended objects in high-resolution SAR images. IEEE Trans. GRS. **43**(4), 833–842 (2005)
19. D. Blacknell, Contextual information in SAR target detection. IEE Proc. Radar, Sonar and Navig. **148**(1), 41–47 (2001)
20. G. Gao, G. Kuang, Q. Zhang, D. Li, Fast detecting and locating groups of targets in high-resolution SAR images. Pattern Recogn. **40**, 1378–1384 (2007)
21. L.M. Novak et al., The automatic target-recognition system in SAIP. Lincoln Lab. J. **10**(2), 187–202 (1997)
22. A. Farrouki, M. Barkat, Automatic censoring CFAR variability for nonhomogeneous environments. IEE Proc. Radar, Sonar and Navig. **152**(1), 43–51 (2005)
23. J.S. Salazar II, *Detection Schemes for Synthetic Aperture Radar Imagery Based On a Beta Prime Statistical Model.* Doctor's dissertation, The New Mexico University, 1999, Chap. 5
24. R.A. English et al., *Development of an ATR Workbench for SAR Imagery*. Technical Report, DRDC, Ottawa, 2005
25. S. Cimmino, G. Franceschetti, A. Iodice, Efficient spotlight SAR raw signal simulation of extended scenes. IEEE Trans. Geosci. Remote Sens. **41**(10), 478–489 (2003)
26. H.H. Arsenault, G. April, Properties of speckle integrated with a finite aperture and logarithmically transformed. J. Opt. Soc. Am. **66**(11), 1160–1163 (1976)
27. K.D. Ward, Compound representation of high resolution sea clutter. Electron. Lett. **7**, 561–565 (1981)
28. A.C. Frery, H.J. Muller, C.C.F. Yanasse, S.J.S. Sant'Anna, A model for extremely heterogeneous clutter. IEEE Trans. GRS **35**(3), 648–659 (1997)
29. G. Moser et al., SAR amplitude probability density function estimation based on a generalized Gaussian scattering model. SPIE **5573**, 307–318 (2004)
30. G. Moser, J. Zerubia, S.B. Serpico, SAR amplitude probability density function estimation based on a generalized Gaussian model. IEEE Trans. Image Process. **15**(6), 1429–1442 (2006)
31. R.O. Duda, P.E. Hart, D.G. Stork, *Pattern Classification*, 2nd edn. (Wiley, New York, 2001)
32. G. Moser, J. Zerubia, S.B. Serpico, Dictionary-based stochastic expectation–maximization for SAR amplitude probability density function estimation. IEEE Trans. Geosci. Remote Sens. **44**(1) (2006)

33. R.A. Rendner, H.F. Walker, Mixture densities, maximum likelihood, and the EM algorithm. SIAM REV **26**(2), 195–239 (1984)
34. I.R. Joughin et al., Maximum likelihood estimation of K distribution parameters for SAR data. IEEE Trans. GRS **31**(5), 989–999 (1993)
35. M.S. Greco, F. Gini, Statistical analysis of high-resolution SAR ground clutter data. IEEE Trans. Geosci. Remote Sens. **45**(3), 566–575 (2007)
36. M.D. DcVore, J.A. O'Sullivan, Statistical assessment of model fit for synthetic aperture radar data. SPIE **4382**, 379–388 (2001)
37. V. Anastassopoulos, High resolution radar clutter statistics. IEEE Trans. AES **35**(1), 43–59 (1999)
38. H. Akaike, Information theory and an extension of maximum likelihood principle, in *2nd International Symposium on Information Theory*, ed. by B.N. Petrov, F. Csaki (Akademiai Kiado, Budapest, 1973), pp. 267–281
39. J.B. Billingsley, A. Farina, F. Gini, M.V. Greco, L. Verrazzani, Statistical analyses of measured radar ground clutter data. IEEE Trans. Aerospace Electronic Systems **35**(3), 579–593 (1999)
40. Y. Delignon, W. Pieczynski, Modelling non-Rayleigh speckle distribution in SAR images. IEEE Trans. GRS. **40**(6), 1430–1435 (2002)
41. A.P. Blake et al., High resolution SAR clutter textural analysis and simulation. SPIE **2584**, 101–108 (1995)
42. G. Gao et al., The multiplicative noise analysis of SAR images. Signal Process. (China) **11**(3), 178–196 (2006). (in Chinese)
43. R.B. D'Agostino, E.S. Pearson, Tests for departure from normality. Empirical results for the distributions of b_2 and $\sqrt{b}_1$. Biometrika **60**(3), 613–622 (1973)
44. D. Blacknell et al., Estimators and distributions in single and multi-look polarimetric and interferometric data, in *IGRASS'94* (IEEE,1994), pp. 8–12
45. E. Ertin, L.C. Potter, Polarimetric classification of scattering centers using M-ary Bayesian decision rules. IEEE Trans. AES **36**(3), 738–749 (2000)
46. I.R. Joughin, D.P. Winebrenner, D.B. Percival, Probability density functions for multilook polarimetric signatures. IEEE Trans. Geosci. Remote Sens. **32**(3), 562–574 (1994)
47. J.S. Lee, K.W. Hoppel, S.A. Mango, A.R. Miller, Inensity and phase statistics of multilook polarimetric and interferometric SAR imagery. IEEE Trans. Geosci. Remote Sens. **32**(5), 1017–1028 (1994)
48. Q. Jiang et al., Ship detection in RADARSAT SAR imagery using PNN-Model, in *Proceedings of ADRO Symposium'98* (1998)
49. E. Parzen, On estimation of probability density function and mode. Signal Process. **33**, 267–281 (1962)
50. C.M. Bishop, *Neural Networks for Pattern Recognition*, 2nd edn. (Oxford University Press, Oxford, U.K., 1996)
51. J. Bruzzone, M. Marconcini, U. Wegmuller, A. Wiesmann, An advanced system for the automatic classification of multitemporal SAR images. IEEE Trans. Geosci. Remote Sens. **42**(6), 1321–1334 (2004)
52. P. Mantero et al., Partially supervised classification of remote sensing images using SVM-based probability density estimation. IEEE Trans. Geosci. Remote Sens. **43**(3), 559–570 (2005)
53. V.N. Vapnik, *Statistical Learning Theory* (Wiley, New York, 1998)
54. J. Weston, A. Gammerman, M. Stitson, V. Vapnik, V. Vovk, C. Watkins, Support vector density estimation, in *Advances in kernel methods support vector learning*, ed. by B. Scholkopf, C.J.C. Burges, A.J. Smola (MIT Press, Cambridge, MA, 1999), pp. 293–306
55. E.E. Kuruoglu, J. Zerubia, Modeling SAR images with a generalization of the Rayleigh distribution. IEEE Trans. Image Process. **13**(4), 527–533 (2004)
56. J.W. Goodman, Some fundamental properties of speckle. J. Opt. Soc. Amer. **66**, 1145–1150 (1977)
57. A. Papoulis, *Probability, random variables, and stochastic processed*, 3rd edn. (McGraw Hill, New York, 1991)

58. L.M. Kaplan, Analysis of multiplicative speckle models for template-based SAR ATR. IEEE Trans. AES **37**(4), 1424–1432 (2001)
59. C.L. Martinez et al., Polarimetric SAR speckle noise model. IEEE Trans. GRS **41**(10), 2232–2242 (2003)
60. L.P. Fred, Texture and speckle in high resolution synthetic aperture radar clutter. IEEE Trans. GRS **31**(1), 192–203 (1993)
61. H. Xie, L.E. Pierce, F.T. Ulaby, Statistical properties of logarithmically transformed speckle. IEEE Trans. GRS **40**(3), 721–727 (2002)
62. M. Thr, K.C. Chin, J.W. Goodman, When is speckle noise multiplicative. Appl. Opt. **21**, 1157–1159 (1982)
63. J.W. Goodman, *Statistical properties of laser speckle patterns, laser speckle and related phenomena* (Springer Verlag, Heidelberg, Germany, 1975), pp. 9–75
64. E. Jakeman, P.N. Pusey, A model for non-Rayleigh sea echo. IEEE Trans. Antennas Propagat. **AP-24**(6), 806–814 (1976)
65. C.J. Oliver, Correlated K-distribution scattering model. Opt. Acta **32**, 1515–1547 (1985)
66. S.H. Yueh, J.A. Kong, K distribution and polarimetric terrain radar clutter. J. Electromagn. Waves Applicat. **3**(8), 747–768 (1989)
67. Y. Delignon, R. Garello, A. Hillion, Statistical modeling of ocean SAR images. Proc. IEE Radar Sonar Navig. **144**, 348–354 (1997)
68. R.S. Raghavan, A method for estimating parameters of K-distributed clutter. IEEE Trans. AES **27**(2), 238–246 (1991)
69. D. Blacknell, Comparison of parameter estimators for K-distribution. IEE Proc.-Radar Sonar Navig. **141**(1), 45–52 (1994)
70. C.J. Oliver, A model for non-Rayleigh scattering statistics. Opt. Acta **31**(6), 701–722 (1984)
71. T. Eltoft, K.A. Hogda, Non-Gaussian signal statistics in ocean SAR imagery. IEEE Trans. Geosci. Remote Sens. **36**(2), 562–575 (1998)
72. J. Jao. Amplitude distribution of composite terrain radar clutter and the K distribution. IEEE Trans. Antennas Propagat. **AP-32**(10), 1049–1052 (1984)
73. J.S. Lee et al., Intensity and phase statistics of multilook polarimetric and interferometric SAR imagery. IEEE Trans. Geosci. Remote Sens. **32**(6), 1017–1027 (1994)
74. R. Barakat, Direct derivation of intensity and phase statistics of speckle produced by a weak scatterer from the random sinusoid model. J. Opt. Soc. Amer. **71**(1), 86–90 (1981)
75. Y. Delignon, A. Marzouki, W. Pieczynski, Estimation of generalized mixtures and its application to image segmentation. IEEE Trans. Image Process. **6**(10), 1364–1375 (2001)
76. A.C. Frery et al., Alternative distributions for the multiplicative model in SAR images, in *International Geoscience Remote Sensing Symposium*, vol. 1, Florence, Italy (1995), pp. 169–171
77. H.J. Muller, Modeling of extremely heterogeneous radar backscatter, in *IGARSS '97*, vol. 4 (1997), 1603–1605
78. H.J. Muller, R. Pac, G-statistics for scaled SAR data, in *Proceedings IEEE Geoscience and Remote Sensing Symposium*, vol. 2 (1999), pp. 1297–1299
79. J.S. Salazar II et al., Statistical modeling of target and clutter in single-look non-polorimetric SAR imagery, in *International Conference Signal and Image Processing*, Las Vegas, USA (1998)
80. B. Jorgensen, *Statistical Properties of the Generalized Inverse Gaussian Distribution.* Lecture Notes in Statistics, 9 (Springer-Verlag, New York, 1982)
81. T. Eltoft, Modeling the amplitude statistics of ultrasonic images. IEEE Trans. Med. Imaging **25**(2), 229–240 (2006)
82. T. Eltoft, The Rician inverse Gaussian distribution: a new model for non-Rayleigh signal amplitude statistics. IEEE Trans. Image Processing **14**(11), 1722–1735 (2005)
83. T. Eltoft, A new model for the amplitude statistics of SAR imagery, in *Proceedings of IGARSS*, vol. III, July 2003, pp. 1993–1995
84. V. Anastassopoulos et al., A generalized compound model for radar clutter, in *IEEE 1994 National Radar Conference*, Atlanta, 1994, pp. 41–45

85. V. Anastassopoulos et al., A new clutter model for SAR images, in *International Conference on Applications of Photonic Technology, ICAPT '94*, Toronto, 1994, pp. 21–23
86. V. Anastassopoulos et al., High resolution radar clutter classification, in *IEEE International Radar Conference*, Washington, DC, 1995, pp. 8–11
87. C.J. Nikias et al., *Signal Processing with Alpha-Stable Distributions and Applications* (Wiley, New York, 1995)
88. R.D. Pierce, RCS characterization using the alpha-stable distribution, in *IEEE 1996 National Radar Conference* (1996), pp. 394–419
89. R. Kappor et al., UWB radar detection of targets in foliage using alpha-stable clutter models. IEEE Trans. AES **35**(3), 819–833 (1999)
90. E.E. Kuruoglu, Density parameter estimation of skewed alpha-stable distributions. IEEE Trans. Signal Process. **49**(10), 2192–2201 (2001)
91. E.E. Kuruoglu, J. Zerubia, Skewed α-stable distributions for modeling textures. Pattern Recognit. Lett. **24**, 339–348 (2003)
92. A. Banerjee, P. Burlina, R. Chellappa, Adaptive target detection in foliage-penetrating SAR images using alpha-stable models. IEEE Trans. Image Process. **8**(12), 1823–1831 (1999)
93. E.E. Kuruoglu et al., Modeling SAR images with a generalization of the Rayleigh distribution, in *The 23th Annual Asilomar Conference on Signals, Systems and Computers*, vol. 1 (2000), pp. 224–228
94. E.E. Kuruoglu et al., Approximation of alpha-stable probability densities using finite mixtures of Gaussian, in *EUSIPCO'98 European Signal Processing Conference*, Rhodes, Greece (1998), pp. 989–992
95. E. Conte, M. Longo, Characterization of radar clutter as a spherically invariant random process, in *IEE Proceedings F-Communications, Radar and Signal Processing*, vol. 134 (Apr. 1987), pp. 191–197
96. S.F. George, *The Detection of Nonfluctuating Targets in Log-Normal Clutter*, NRL Report 6796 (1968)
97. S. Kuttikkad, R. Chellappa, Non-Gaussian CFAR techniques for target detection in high resolution SAR images, in *Proceedings of ICIP-94* (1994), pp. 910–914
98. K. Fukunaga, *Introduction to Statistical Pattern Recognition*, 2nd edn. (Academic, Orlando, FL, 1990)
99. F.T. Ulaby et al., Textural information in SAR images. IEEE Trans. GRS **24**, 235–245 (1986)
100. R. Dana, D. Knepp, The impact of strong scintillation on space based radar design II: noncoherent detection. IEEE Trans. Aerosp. Electron. Syst. **AES-22**(1), 34–36 (1986)
101. M.D. DeVore et al., ATR performance of a Rician model for SAR images. *SPIE* (2000), p. 4050
102. A.P. Blake et al., High resolution SAR clutter textural analysis. *IEE Colloquium on Recent Developments in Radar and Sonar Imaging Systems: What Next?* 1995, UK, 10/1–10/9
103. D. Borghys, *Interpretation and Registration of High-Resolution Polarimetric SAR Images*. ENSTE 031, Paris, 2001
104. D. Blacknell, A mixture distribution model for correlated SAR clutter. SPIE **2958**, 38–49 (1996)
105. D. Blacknell, Target detection in correlated SAR clutter. IEE Proc-RSN **147**(1), 9–16 (2000)
106. D.A. Shnidman, Generalized radar clutter model. IEEE Trans. AES **35**(3), 857–865 (1999)
107. L.P. Fred, Texture and speckle in high resolution synthetic aperture radar clutter. IEEE Trans. Remote Sens. **31**(1), 192–203 (1993)
108. A. Lopes et al., Statistical distribution and texture in multilook and complex SAR images, in *International Geoscience and Remote Sensing Symposium*, vol. 3, Washington, DC, 1990, pp. 2427–2430

Chapter 2
Statistical Modeling of Single-Channel SAR Images

2.1 Modeling SAR Images Based on a Generalized Gamma Distribution for Texture Component

There has been a growing interest in SAR image interpretation for a variety of civilian and military applications, such as terrain classification, land or sea monitoring, target detection and recognition, etc. Presently, one of major strategies of SAR image processing is to use the classical methods of statistical pattern recognition, where it is crucial to develop precise models for the statistics of the pixel amplitudes or intensities [1].

Studies on statistical models of SAR images with different terrain types have been carried out over the last couple of decades. In summary, three main solutions, i.e., parametric models [2–12], non-parametric models [13–15] and mixture models [16–20], can be employed for this purpose. Specifically, as a modeling way of the data-driven, non-parametric approaches, e.g., probabilistic neural network (PNN) [15] and standard Parzen window estimator [13, 14], have proved to be effective estimated tools and suitable for estimating the complex or unknown probability density function (PDF). However, as vast computational load and numerous data involved, non-parametric approaches are limited in the various applications. Consequently, parametric models turn out to be intensively investigated in the literature [21]. The process of parametric modeling can be described in brief as to choose the appropriate one from several analytical statistical distributions for the given image. Practically, it is not sufficient for fitting a large-scale scene by parametric models due to the complex contexts and numerous terrain types included [22]. The mixture models, combining two or more parametric models, provide a creative idea for this problem. Several alternatives, such as Gamma mixture model [16], the finite mixture models [18], etc., are given. Unfortunately, the estimates of parameters are a hard task for this kind of models. Similarly with the time-consuming of non-parametric approaches, a common strategy, for instance, using expectation-maximization (EM) algorithm [19, 20], often tends to be tediously iterative.

G. Gao, *Characterization of SAR Clutter and Its Applications to Land and Ocean Observations*, https://doi.org/10.1007/978-981-13-1020-1_2

Focusing on parametric models, the product model [1], expressed as the product of an underlying radar cross section (RCS) with an uncorrelated multiplicative speckle noise one, has been widely and successfully used in conducting the analytical mathematic formulas to describe the statistical properties of SAR data. Some studies have been presented based on this model in the past several decades, such as Γ, $\mathcal{K}$ etc. Among these distributions, an important contribution is the derivation of the famous $\mathcal{G}^0$ distribution by Frery [7], under the consideration of discriminating types of homogeneous, heterogeneous and extremely heterogeneous terrains. This distribution encompasses most of the existing models, mainly because the reciprocal of a Gamma (also called the inverse Gamma) distribution [7] can match the RCS components of different terrain types. Recently, Li et al. [23] proposed an empirical generalized Gamma distribution GΓD to estimate the PDFs of SAR data. Although this distribution is a purely mathematical model and has no relation to the physics of wave scattering [24], many well-known distributions, such as Rayleigh, exponential, Nakagami, Gamma, Weibull, log-normal and inverse Gamma distributions, are particular cases of this GΓD [23].

Motivated by aforementioned characteristic, this chapter is first devoted to developing a novel statistical model (denoted simply as $G\Gamma\Gamma$) to model different types of clutter with respect to pixel amplitude or intensity in high-resolution SAR images. The proposed model arises using GΓD proposed to describe the RCS components of amplitude or intensity return based on the product model. It could be proven theoretically that the new model has the $\mathcal{K}$ distribution for heterogeneous clutter as well as the $\mathcal{G}^0$ distribution for extremely heterogeneous clutter as special cases. Furthermore, using the second-kind statistics theory developed by Nicolas [25, 26], which relies on the Mellin transform, i.e., "method-of-log-cumulants" (MoLC), we derive the parameter estimators of the new distribution model.

2.1.1 *The Proposed* G$\Gamma\Gamma$ *Model*

2.1.1.1 The GΓD and the Product Models

The generalized gamma distribution is recently defined as [23].

$$p_X(x) = \frac{|\nu|\kappa^\kappa}{\sigma\Gamma(\kappa)}\left(\frac{x}{\sigma}\right)^{\kappa\nu-1}\exp\left\{-\kappa\left(\frac{x}{\sigma}\right)^\nu\right\}, \quad \sigma, |\nu|, \kappa, x > 0 \tag{2.1}$$

where ν, κ and σ are the power, shape and scale parameters, respectively. $\Gamma(\bullet)$ represents the Gamma function. The PDF characterized by (2.1) is shown as an empirical model. It is an ideal alternative for modeling the RCS components of SAR intensity or amplitude return, owing to that both the Gamma distribution (describing the heterogeneous clutter) and the inverse Gamma one (describing the extremely heterogeneous terrains) can be viewed as particular cases of this general model [23], and correspond to $\nu = 1$ and $\nu = -1$, respectively.

The product model [1] has been testified that it is valid for conducting statistical models of amplitude or intensity statistics of SAR data. Its expression is given by

$$Z = X \cdot Y \tag{2.2}$$

where X and Y indicates the backscattering RCS component and speckle noise one, respectively, Z denotes the observed intensity or amplitude of SAR data. To make a clear distinction, a similar strategy with literature [7], i.e., the subscripts "I" and "A", will be used separately hereafter for the intensity and amplitude cases. Considering the case of the intensity, it is widely accepted that multilook intensity speckle noise component obeys the Gamma distribution [7] with unitary mean, whose PDF is expressed by

$$p_{Y_I}(y) = \frac{n^n}{\Gamma(n)} y^{n-1} \exp(-ny), \quad y, n > 0 \tag{2.3}$$

where n is the number of looks.

2.1.1.2 The $G\Gamma\Gamma$ Model

Herein, within the structure of the product model shown in (2.2), combining (2.3) and (2.1), the PDF of Z_I can be derived as

$$\begin{aligned} p_{Z_I}(z) &= \int_0^\infty \frac{1}{x} \cdot p_{X_I}(x) \cdot p_{Y_I}\left(\frac{z}{x}\right) dx \\ &= \frac{|v| \kappa^\kappa n^n}{\sigma^{\kappa v} \Gamma(\kappa) \Gamma(n)} z^{n-1} \int_0^\infty \kappa^{v-n-1} \exp\left\{-\kappa\left(\frac{x}{\sigma}\right)^v - n\left(\frac{z}{x}\right)\right\} dx, \\ & \quad \sigma, \kappa, n, z > 0, v \neq 0 \end{aligned} \tag{2.4}$$

Its moments of m order turn out to be (see Appendix 2-A):

$$E\left(z_I^m\right) = \frac{\Gamma(m+n)}{\Gamma(\kappa)\Gamma(n)} \Gamma\left(\kappa + \frac{m}{v}\right) \left(\frac{\sigma \kappa^{-1/v}}{n}\right)^m \tag{2.5}$$

In order to facilitate the numerical integral in (2.4), a transform, $x = \tan(\theta)$, leads to that (2.4) could be rewritten as

$$p_{Z_I}(z) = \frac{|v| \kappa^\kappa n^n}{\sigma^{\kappa v} \Gamma(\kappa) \Gamma(n)} z^{n-1}.$$

$$\int_0^{\frac{\pi}{2}} \sec^2(\theta)\tan^{\kappa v-n-1}(\theta)\exp\left\{-\kappa\left(\frac{\tan(\theta)}{\sigma}\right)^v - n\left(\frac{z}{\tan(\theta)}\right)\right\}d\theta,$$

$$\sigma, \kappa, n, z > 0, v \neq 0 \tag{2.6}$$

Letting Z_A denote as the amplitude return, via the relationship of $p_{Z_A}(z) = 2zp_{Z_I}(z^2)$, and thus the PDF of Z_A is characterized by

$$p_{Z_A}(z) = \frac{2|v|\kappa^\kappa n^n}{\sigma^{\kappa v}\Gamma(\kappa)\Gamma(n)} z^{2n-1}.$$

$$\int_0^{\frac{\pi}{2}} \sec^2(\theta)\tan^{\kappa v-n-1}(\theta)\exp\left\{-\kappa\left(\frac{\tan(\theta)}{\sigma}\right)^v - n\left(\frac{z^2}{\tan(\theta)}\right)\right\}d\theta,$$

$$\sigma, \kappa, n, z > 0, v \neq 0 \tag{2.7}$$

We refer to this distribution characterized by (2.6) or (2.7) as the generalized Gamma Gamma distribution, abbreviated as $G\Gamma\Gamma$ distribution. Specifically, we call the $G\Gamma\Gamma_I$ distribution and the $G\Gamma\Gamma_A$ distribution, correspond to (2.6) and (2.7), respectively, to distinct the intensity statistic as well as the amplitude statistic. Figure 2.1 gives the plots of the $G\Gamma\Gamma_A$ distribution with respect to the various parameters.

Additionally, it can be proven (see Appendix 2-B) theoretically that the $G\Gamma\Gamma$ is degenerating to the well-known $\mathcal{G}^0$ and $\mathcal{K}$ on the conditions of $v = -1$ and $v = 1$, respectively. This property stated means that the proposed $G\Gamma\Gamma$ distribution encompasses the modeling abilities of $\mathcal{G}^0$ and $\mathcal{K}$ whilst extending them to enable the modeling of the clutter areas with more widely varying degrees of homogeneity.

2.1.2 Parameter Estimator of the $G\Gamma\Gamma$ Model Based on MoLC

2.1.2.1 The Log-Cumulants of $G\Gamma\Gamma$

The MoLC has been proposed by Nicolas as a parametric PDF estimation technique for a function defined over $\mathbb{R}^+$. MoLC is based on a second kind statistics by applying the Mellin transform, instead of the Fourier and Laplace transforms. Given a positive-valued random variable X with the PDF $p_X(x)$, the second-kind first characteristic function is defined as the Mellin transform of $p_X(x)$:

$$\phi_X(s) = \mathcal{M}[p_X(x)](s) = \int_0^\infty x^{s-1}p_X(x)dx \tag{2.8}$$

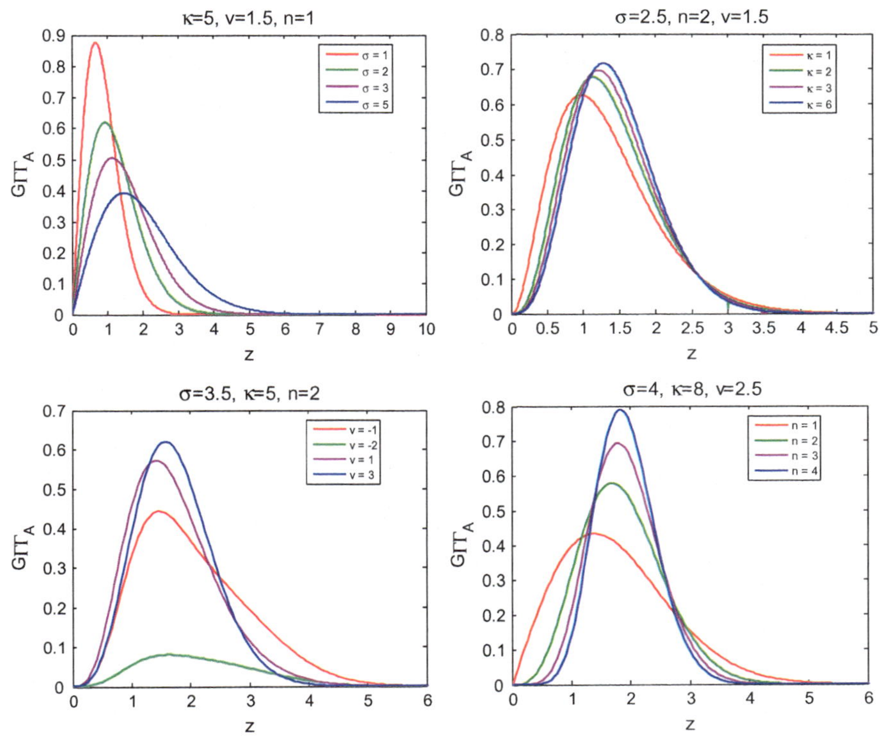

Fig. 2.1 Plots of $G\Gamma\Gamma_A$ versus different parameters

where $\mathcal{M}$ is the Mellin transform operator. Subsequently, the second characteristic function of the second kind is computed as the natural logarithm of $\phi_X(s)$.

$$\varphi_X(s) = \log(\phi_X(s)) \tag{2.9}$$

The rth order derivative of $\varphi_X(s)$ at $s = 1$ is the second kind cumulants (also named log-cumulants) of order r, i.e.,

$$\tilde{c}_r = \left. \frac{d^r \varphi_X(s)}{ds^r} \right|_{s=1} \tag{2.10}$$

Thanks to the relation between the Mellin transform and the moments of random variable, i.e., $\phi_X(s) = E\left(X^{s-1}\right)$. Hence, via (2.5), the second-kind first and second characteristic functions of the $G\Gamma\Gamma$ distribution are given, respectively, by the following equations.

$$\begin{cases} \phi_{G\Gamma\Gamma}(s) = \frac{\Gamma(s-1+n)}{\Gamma(\kappa)\Gamma(n)}\Gamma\left(\kappa + \frac{s-1}{v}\right)\left(\frac{\sigma}{n\kappa^{\frac{1}{v}}}\right)^{s-1} \\ \varphi_{G\Gamma\Gamma}(s) = \log(\Gamma(s-1+n)) + \log\left(\Gamma\left(\kappa + \frac{s-1}{v}\right)\right) \\ \qquad - \log(\Gamma(\kappa)) - \log(\Gamma(n)) + (s-1)\left(\log(\sigma) - \log(n) - \frac{1}{v}\log(\kappa)\right) \end{cases} \tag{2.11}$$

From (2.10) and (2.11), one can obtain the log-cumulants of the $G\Gamma\Gamma$ distribution expressed by

$$\begin{cases} \tilde{c}_1 = \Psi(n) + \log(\sigma/n) + (\Psi(\kappa) - \log(\kappa))/v \\ \tilde{c}_r = \Psi(r-1, n) + \Psi(r-1, \kappa)/v^r, \quad r \geq 2 \end{cases} \tag{2.12}$$

where $\Psi(\cdot)$ represents the digamma function (i.e., the logarithmic derivative of the Gamma function), and $\Psi(r, \cdot)$ is the rth order polygamma function (i.e., the kth order derivative of the digamma function).

Given a sample set $\{z_i\}$, $i \in [1, N]$, the estimates of log-cumulants can be acquired directly by

$$\begin{cases} \hat{\tilde{c}}_1 = \frac{1}{N}\sum_{i=1}^{N}[\ln(z_i)] \\ \hat{\tilde{c}}_r = \frac{1}{N}\sum_{i=1}^{N}\left[\left(\ln(z_i) - \hat{\tilde{c}}_1\right)^r\right], \quad r \geq 2 \end{cases} \tag{2.13}$$

2.1.2.2 The Estimates of Parameters in $G\Gamma\Gamma$

Paying attention to (2.12), we stress that $\tilde{c}_r$, $r \geq 2$ do not contain σ, thus allowing us to divide the parameter estimates of the $G\Gamma\Gamma$ distribution into three distinct stages.

First, based on (2.12), the 2th and 3th order log-cumulants of the $G\Gamma\Gamma$ distribution yield

$$\tilde{c}_2 = \Psi(1, n) + \Psi(1, \kappa)/v^2 \tag{2.14}$$

$$\tilde{c}_3 = \Psi(2, n) + \Psi(2, \kappa)/v^3 \tag{2.15}$$

To isolate v in previous two equations, we take the ratio of $\Psi^3(1, \kappa)/\Psi^2(2, \kappa)$, and thus via (2.14) and (2.15), one arrive at that the estimate $\hat{\kappa}$ of the parameter κ is given by

$$\frac{\Psi^3(1, \hat{\kappa})}{\Psi^2(2, \hat{\kappa})} = \frac{\left(\hat{\tilde{c}}_2 - \Psi(1, n)\right)^3}{\left(\hat{\tilde{c}}_3 - \Psi(2, n)\right)^2} \tag{2.16}$$

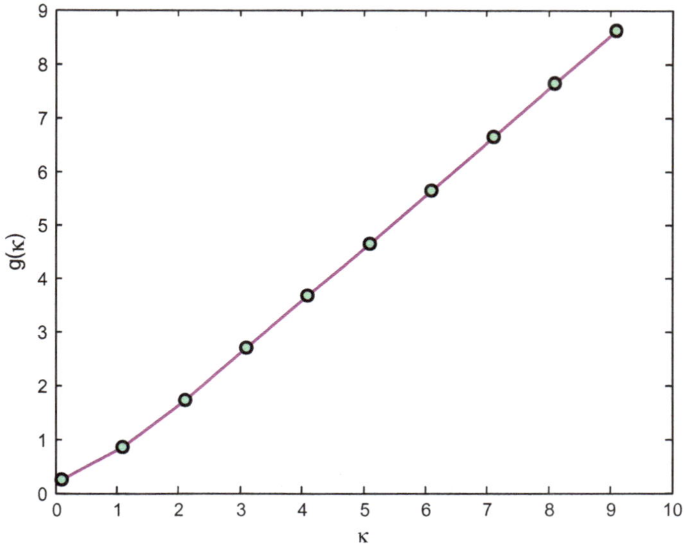

Fig. 2.2 Plot of the function $g(\hat{\kappa})$

As shown in Fig. 2.2, the function on the left-hand-side of (2.16), $g(\hat{\kappa}) = \Psi^3(1, \hat{\kappa})/\Psi^2(2, \hat{\kappa})$, is strictly monotonically increasing. A simple numerical solution, for instance, the f solve function in Matlab or the bisection method [27], can be adopted to obtain the value of $\hat{\kappa}$.

Next, taking $\hat{\kappa}$ into (2.15) leads to that the estimate $\hat{v}$ of the parameter v is

$$\hat{v} = \sqrt[3]{\frac{\Psi(2, \hat{\kappa})}{\hat{\hat{c}}_3 - \Psi(2, n)}} \tag{2.17}$$

Finally, plugging $\hat{v}$ and $\hat{\kappa}$ into the first equation of (2.12), the estimate $\hat{\sigma}$ turns out to be solved.

2.1.3 Experimental Results

The aim of this section is to verify the modeling capability of the proposed $G\Gamma\Gamma$ distribution based on the measured SAR data. As discussed in earlier works [7], we note that $\mathcal{G}^0$ is an accurate model for SAR images over homogeneous, heterogeneous and extremely heterogeneous terrains (like urban areas), and some conclusions [7] have been made that the $\mathcal{G}^0$ distribution outperforms other conventional distributions, such as $\mathcal{K}$, Weibull, Nakagami, etc., in estimating PDFs of given SAR images. On the other hand, by previous analysis, the proposed $G\Gamma\Gamma$ distribution encompasses theoretically the modeling abilities of $\mathcal{G}^0$ as proven in Appendix 2-B. Hence, we only compare the proposed $G\Gamma\Gamma$ distribution with the $\mathcal{G}^0$[1] one in this section.

It should be emphasized that we have tested the $G\Gamma\Gamma$ and $\mathcal{G}^0$ models on various SAR images (e.g., L-band airborne SAR data acquired by the NASA/JPL AIRSAR sensor, SAR data provided by RADARSAT-2 space-borne SAR system operated in C band, TerraSAR-X SAR data of X band and other data provided by Chinese airborne sensors), which involve different bands, platforms, polarimetric modes and spatial resolutions. Consequently, we found that the performances of $G\Gamma\Gamma$ are almost in agreement with that of $\mathcal{G}^0$ when fitting the observed histograms of given SAR images in low-resolution and the homogeneous, heterogeneous regions in high-resolution. In other words, these tested results show that the $G\Gamma\Gamma$ distribution can model all cases the $\mathcal{G}^0$ distribution can, as expected.

Therefore, to better compare the aforementioned two statistical models, we further check them on extremely heterogeneous terrains of high-resolution images. As an example, a large TerraSAR-X spotlight-mode 300 MHz geocoded scene over Beijing of China, acquired on January 31, 2008 at 10:03 UTC, with high-resolution 2 m $\times$ 2.3 m (azimuth $\times$ range) and HH-polarization is used in this chapter as shown in Fig. 2.3. Like other cities in the world, Beijing is known for its well-planned urban expansion. Thus, it remains valid for other cities with similar appearance, although hereafter the tested results on Beijing only are given.

Three representative patches of extremely heterogeneous areas dominated by urban of buildings and indicated by boxes in Fig. 2.3, numbered A-C (with the sizes of 1476 $\times$ 1342, 1868 $\times$ 1848, and 2361 $\times$ 1720 pixels, respectively), are selected as the main regions of this study. Figures 2.4, 2.5 and 2.6 show the comparison of the amplitude histograms of the three selected areas indicated in Fig. 2.3 with the proposed $G\Gamma\Gamma$ and classical $\mathcal{G}^0$ model fits. The estimating results on the linear and logarithmic scales are all provided so that the performance difference in fitting the whole and the tails can be seen clearly.

Before the estimates of other parameters in the proposed $G\Gamma\Gamma$ distribution characterized by (2.6) or (2.7), the number of looks n is replaced by the ENL. Thus, the estimate $\hat{n}$ of n is equal to (2.22) about the scene as shown in Fig. 2.3 by using homogeneous areas. Next, the parameter estimations of each distribution in Figs. 2.4, 2.5 and 2.6 are all accomplished by the estimators based on the MoLC, which are shown at Table 2.1, where the analytical expression of the $\mathcal{G}^0$ distribution is given in (2.26).

In order to quantitatively assess the fitting result, we adopt the Kolmogorov–Smirnov (KS) test [1, 28] as a similarity measurement. The smaller the value of KS measurement obtains, the higher fitting accuracy they have. The KS values of the fitting results shown in Figs. 2.4, 2.5 and 2.6 are compared in Table 2.2. From this table, it can be conclude that the proposed $G\Gamma\Gamma$ distribution well agrees with the given SAR images, which implies the higher precision of fitting using $G\Gamma\Gamma$ distribution than using $\mathcal{G}^0$ distribution.

[1]Throughout this chapter, the estimates of parameters of the $\mathcal{G}^0$ distribution adopt the MoLC, the detailed description is given by Tison et al. [18].

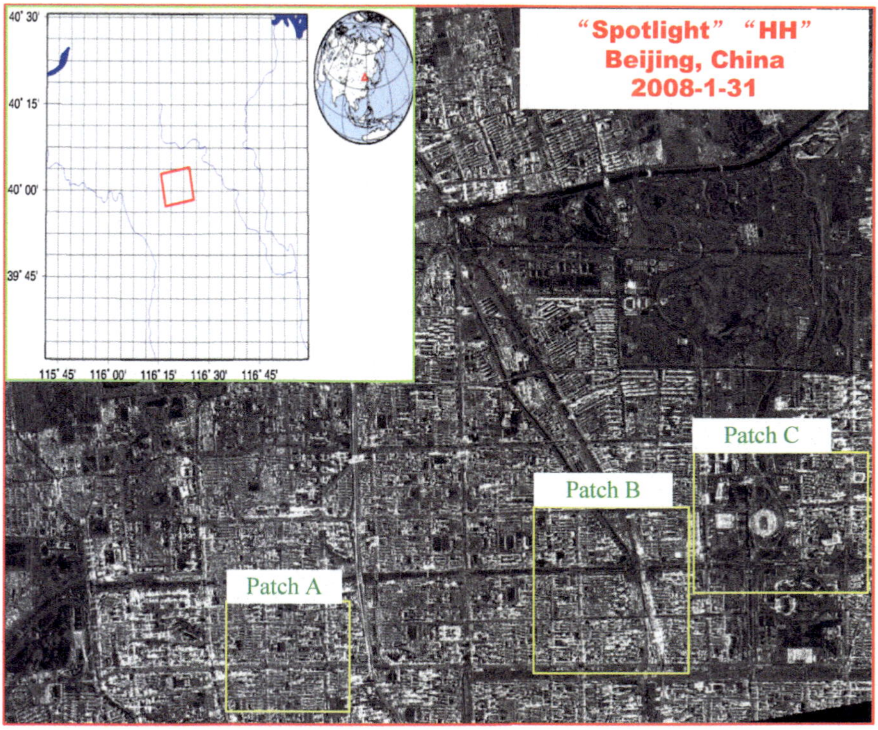

Fig. 2.3 The SAR image of Beijing from the TerraSAR-X

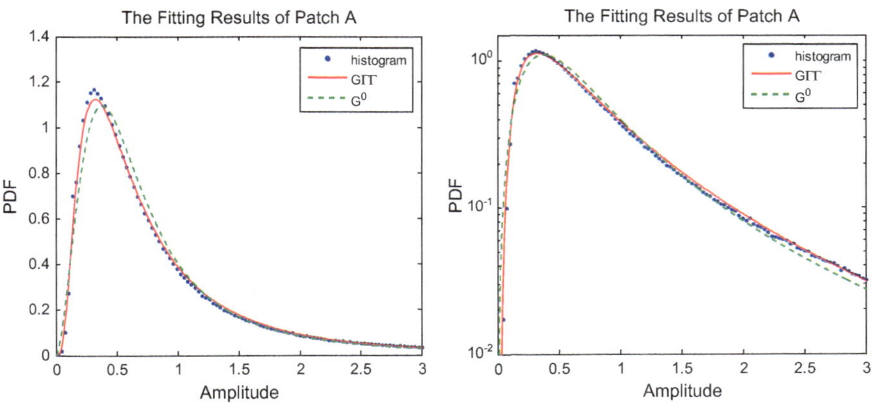

Fig. 2.4 Plots of amplitude histogram for Patch A and of the estimated $G\Gamma\Gamma$ and $\mathcal{G}^0$ PDFs (left: in linear scale, right: in logarithmic scale)

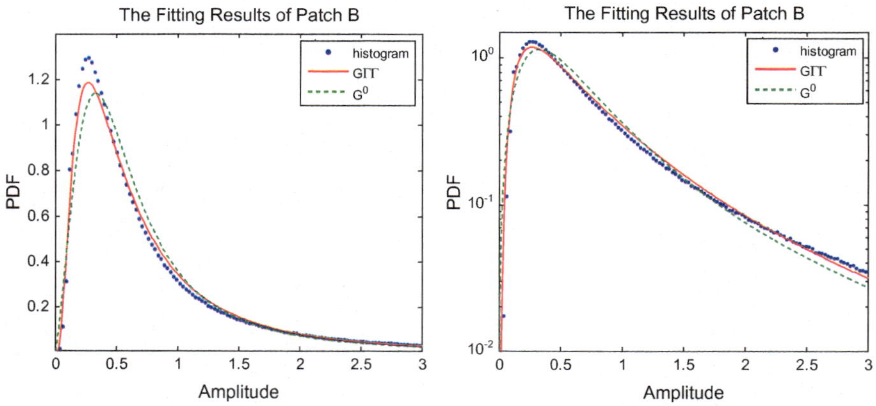

Fig. 2.5 Plots of amplitude histogram for Patch B and of the estimated $G\Gamma\Gamma$ and $\mathcal{G}^0$ PDFs (left: in linear scale, right: in logarithmic scale)

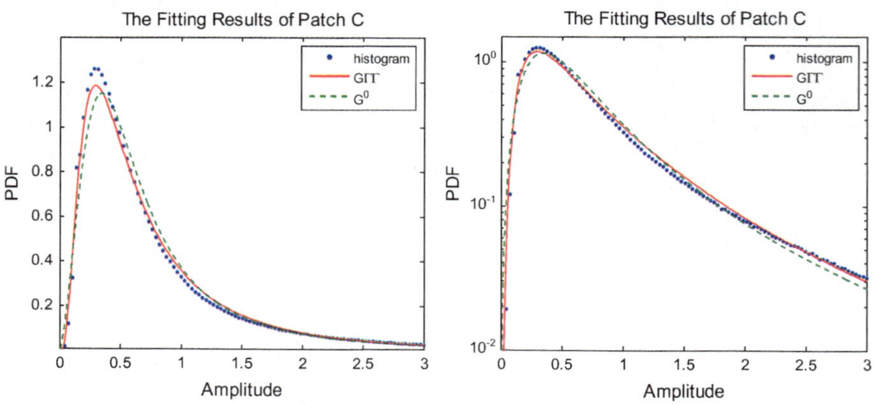

Fig. 2.6 Plots of amplitude histogram for Patch C and of the estimated $G\Gamma\Gamma$ and $\mathcal{G}^0$ PDFs (left: in linear scale, right: in logarithmic scale)

Table 2.1 Parameter estimations of noted areas in Fig. 3

Patches	$G\Gamma\Gamma$ $(\hat{\sigma}, \hat{\kappa}, \hat{v})$	$\mathcal{G}^0$ $(\hat{\alpha}, \hat{\gamma}, \hat{n})$
A	(0.3781, 4.0953, −0.3360)	(−0.9260, 0.3332, 1.2713)
B	(0.3229, 4.3008, −0.2990)	(−0.8379, 0.2584, 1.1484)
C	(0.3239, 3.1509, −0.3744)	(−0.8630, 0.2708, 1.2668)

Table 2.2 The comparison of fitting histogram

Patches	KS values of $G\Gamma\Gamma$	KS values of $\mathcal{G}^0$
A	0.0053	0.0131
B	0.0101	0.0206
C	0.0082	0.0153

2.1.4 Appendix 2-A. The Derivation of mth Order Moments of the GΓΓ Distribution

Via (4), The mth order moment estimation of $G\Gamma\Gamma$ is

$$\begin{aligned} E\left(z_I^m\right) &= \int_0^\infty z^m f_{Z_I}(z)dz \\ &= \int_0^\infty z^m \frac{|v|\kappa^\kappa n^n}{\sigma^{\kappa v}\Gamma(\kappa)\Gamma(n)} z^{n-1} \int_0^\infty x^{\kappa v-n-1} \exp\left\{-\kappa\left(\frac{x}{\sigma}\right)^v - n\left(\frac{z}{x}\right)\right\}dxdz \\ &= \frac{|v|\kappa^\kappa n^n}{\sigma^{\kappa v}\Gamma(\kappa)\Gamma(n)} \int_0^\infty x^{\kappa v-n-1} \exp\left\{-\kappa\left(\frac{x}{\sigma}\right)^v\right\} \int_0^\infty z^{m+n-1} \exp\left\{-n\left(\frac{z}{x}\right)\right\}dzdx \end{aligned} \tag{2.18}$$

Letting $t = nz/x$, then $z = tx/n$ and $dz = \frac{x}{n}dt$. Due to $z : 0 \to \infty$, $t : 0 \to \infty$. Thus, the integral of right-hand side in (2.18) is given by

$$\begin{aligned} &\int_0^\infty z^{m+n-1} \exp\left\{-n\left(\frac{z}{x}\right)\right\}dz \\ &= \int_0^\infty \left(\frac{tx}{n}\right)^{m+n-1} \exp\{-t\}\frac{x}{n}dt \\ &= \left(\frac{x}{n}\right)^{m+n} \int_0^\infty t^{m+n-1} \exp\{-t\}dt \\ &= \left(\frac{x}{n}\right)^{m+n} \Gamma(m+n) \end{aligned} \tag{2.19}$$

Taking (2.19) into (2.18), (2.18) can be simplified as

$$E\left(z_I^m\right) = \frac{|v|\kappa^\kappa n^n}{\sigma^{\kappa v}\Gamma(\kappa)\Gamma(n)} \int_0^\infty x^{\kappa v-n-1} \exp\left\{-\kappa\left(\frac{x}{\sigma}\right)^v\right\}\left(\frac{x}{n}\right)^{m+n}\Gamma(m+n)dx$$
$$= \frac{|v|\kappa^\kappa \Gamma(m+n)}{\sigma^{\kappa v}\Gamma(\kappa)\Gamma(n)n^m} \int_0^\infty x^{\kappa v+m-1} \exp\left\{-\kappa\left(\frac{x}{\sigma}\right)^v\right\}dx \tag{2.20}$$

A change of $t = \kappa\left(\frac{x}{\sigma}\right)^v$ leads to $x = \sigma\left(\frac{t}{\kappa}\right)^{1/v}$ and $dx = \frac{\sigma}{\kappa v}\left(\frac{t}{\kappa}\right)^{1/v-1}dt$, where $x: 0 \to \infty$, then $t: \begin{cases} 0 \to \infty, & v > 0 \\ \infty \to 0, & v < 0 \end{cases}$. Thus, two cases are divided. When $v > 0$, we obtain

$$E\left(z_I^m\right) = \frac{|v|\kappa^\kappa \Gamma(m+n)}{\sigma^{\kappa v}\Gamma(\kappa)\Gamma(n)n^m} \int_0^\infty \sigma^{\kappa v+m-1}\left(\frac{t}{\kappa}\right)^{\frac{\kappa v+m-1}{v}} \exp\{-t\}\frac{\sigma}{\kappa v}\left(\frac{t}{\kappa}\right)^{1/v-1} dt$$
$$= \frac{v\Gamma(m+n)\sigma^m}{v\Gamma(\kappa)\Gamma(n)n^m\kappa^{\frac{m}{v}}} \int_0^\infty t^{\kappa+\frac{m}{v}-1} \exp\{-t\}dt$$
$$= \frac{\Gamma(m+n)}{\Gamma(\kappa)\Gamma(n)}\Gamma\left(\kappa+\frac{m}{v}\right)\left(\frac{\sigma\kappa^{-1/v}}{n}\right)^m \tag{2.21}$$

Oppositely, if $v < 0$, (2.20) is expressed by

$$E\left(z_I^m\right) = \frac{|v|\kappa^\kappa \Gamma(m+n)}{\sigma^{\kappa v}\Gamma(\kappa)\Gamma(n)n^m} \int_\infty^0 \sigma^{\kappa v+m-1}\left(\frac{t}{\kappa}\right)^{\frac{\kappa v+m-1}{v}} \exp\{-t\}\frac{\sigma}{\kappa v}\left(\frac{t}{\kappa}\right)^{1/v-1} dt$$
$$= \frac{-v\Gamma(m+n)\sigma^m}{v\Gamma(\kappa)\Gamma(n)n^m\kappa^{\frac{m}{v}}} \int_\infty^0 t^{\kappa+\frac{m}{v}-1} \exp\{-t\}dt$$
$$= \frac{v\Gamma(m+n)\sigma^m}{v\Gamma(\kappa)\Gamma(n)n^m\kappa^{\frac{m}{v}}} \int_0^\infty t^{\kappa+\frac{m}{v}-1} \exp\{-t\}dt$$
$$= \frac{\Gamma(m+n)}{\Gamma(\kappa)\Gamma(n)}\Gamma\left(\kappa+\frac{m}{v}\right)\left(\frac{\sigma\kappa^{-1/v}}{n}\right)^m \tag{2.22}$$

The (2.21) or (2.22) are completely identical to the expression shown in (2.5).

2.1.5 Appendix 2-B. Proof of the Relationship Between Distributions

2.1.5.1 Proof of $G\Gamma\Gamma \overset{V=-1}{\rightarrow} \mathcal{G}^0$

On the condition of $v = -1$, (2.4) turns out to be

$$f_{Z_I}(z) = \frac{\kappa^\kappa n^n}{\sigma^{-\kappa}\Gamma(\kappa)\Gamma(n)} z^{n-1} \int\limits_0^\infty x^{-\kappa-n-1} \exp\left\{-\kappa\frac{\sigma}{x} - n\left(\frac{z}{x}\right)\right\} dx \tag{2.23}$$

According to the integral formula $\int\limits_0^\infty x^{a-1} \exp\{-bx\} dx = \frac{1}{b^a}\Gamma(a)$, $\mathrm{Re}a > 0$, $\mathrm{Re}b > 0$ [29] and making a variable change of $t = 1/x$, we yield

$$\begin{aligned} f_{Z_I}(z) &= \frac{\kappa^\kappa n^n}{\sigma^{-\kappa}\Gamma(\kappa)\Gamma(n)} z^{n-1} \int\limits_\infty^0 -t^{-2} \cdot t^{\kappa+n+1} \exp\{-(\kappa\sigma + nz)t\} dt \\ &= \frac{\kappa^\kappa n^n}{\sigma^{-\kappa}\Gamma(\kappa)\Gamma(n)} z^{n-1} \int\limits_0^\infty t^{\kappa+n-1} \exp\{-(\kappa\sigma + nz)t\} dt \\ &= \frac{\kappa^\kappa n^n}{\sigma^{-\kappa}\Gamma(\kappa)\Gamma(n)} z^{n-1} \frac{1}{(\kappa\sigma + nz)^{\kappa+n}} \Gamma(\kappa + n) \end{aligned} \tag{2.24}$$

Letting $\kappa = -\alpha$ and $\gamma = \kappa\sigma$, the well-known intensity $\mathcal{G}^0$ distribution [7] is obtained, i.e.,

$$f_{Z_I}(z) = \frac{n^n \Gamma(n-\alpha) z^{n-1}}{\gamma^\alpha \Gamma(n)\Gamma(-\alpha)(\gamma + nz)^{n-\alpha}}, \quad -\alpha, \gamma, n, z > 0 \tag{2.25}$$

2.1.5.2 Proof of $G\Gamma\Gamma \overset{V=1}{\rightarrow} \mathcal{K}$

Similarly, when $v = 1$, (2.4) can be rewritten as

$$f_{Z_I}(z) = \frac{\kappa^\kappa n^n}{\sigma^{\kappa}\Gamma(\kappa)\Gamma(n)} z^{n-1} \int\limits_0^\infty x^{\kappa-n-1} \exp\left\{-\kappa\frac{x}{\sigma} - n\left(\frac{z}{x}\right)\right\} dx \tag{2.26}$$

Utilizing the integral formula $\int\limits_0^\infty x^{a-1} \exp\left(-\frac{c}{x} - bx\right) dx = 2\left(\frac{c}{b}\right)^{a/2} K_a\left(2\sqrt{cb}\right)$, $\mathrm{Re}b > 0, \mathrm{Re}c > 0$ [34], we further obtain

$$f_{Z_I}(z) = \frac{\kappa^\kappa n^n}{\sigma^\kappa \Gamma(\kappa)\Gamma(n)} z^{n-1} 2\left(\frac{nz\sigma}{\kappa}\right)^{\frac{\kappa-n}{2}} K_{\kappa-n}\left(2\sqrt{nz\kappa/\sigma}\right) \tag{2.27}$$

By the replacement of variable, i.e., $\lambda = \kappa/\sigma$ and $\alpha = \kappa$, we arrive at the $\mathcal{K}$ distribution [14]:

$$f_{Z_I}(z) = \frac{2\lambda n}{\Gamma(\alpha)\Gamma(n)} (\lambda n z)^{\frac{\alpha+n}{2}-1} K_{\alpha-n}\left(2\sqrt{n\lambda z}\right) \quad \alpha, \lambda, n, z > 0 \tag{2.28}$$

where $K_{\alpha-n}(\cdot)$ is the second type modified Bessel function with order $\alpha - n$.

2.2 Scheme for Characterizing Clutter Statistics in SAR Amplitude Images by Combining Two Parametric Models

In empirical models, the distributions are constructed from a purely mathematical perspective, regardless of the physical mechanism responsible for the backscattering of the radar clutter. Some well-known examples of these distributions are the *log-normal* (LN) [1], *Weibull* [30], *Fisher* [27] (completely identical to the $\mathcal{G}^0$), and GΓD distributions [23]. The GΓD, in particular, has proved to be a good candidate for describing both land and sea clutter data in SAR images. In the case of land clutter, the GΓD has been used for modeling many types of ground clutter in high-resolution SAR images, and it has exhibited better performance than many parametric distributions (for example, the *Nakagami*, *Weibull*, $\mathcal{K}$, *Fisher*, GGR, and GΓR) in most cases for fitting real data [23]. In the case of sea clutter, [31–33] reported results of fitting single-look and multi-look SAR data having different polarizations and sea states and showed that better fitting accuracy could be obtained by the GΓD than by many classic models, including the *Gaussian*, *Rayleigh*, *gamma*, LN, *Weibull*, $\mathcal{K}$, and GK.

Parameter estimation is the first step when using a parametric model in practice. The MoLC is a parameter estimation technique that is well-suited to most SAR-specific statistical models. It is a good option when alternatives, such as the maximum likelihood (ML) or the method of moments (MoM), fail to provide acceptable estimators [34]. Estimators for the GΓD based on the MoLC have been expressed in [23]. A further parameter estimation scheme, covering a wider range of applicable conditions than the original MoLC-based estimators, was presented in Gao et al.[33]. In addition, it is worth noting that an empirical model named $\mathcal{G}_{\text{AO}}$ was proposed in [35]. The experimental results tested on satellite SAR data of sea clutter validated the soundness of the model and demonstrated that the $\mathcal{G}_{\text{AO}}$ model is superior to the commonly-used $\mathcal{K}$ distribution for fitting sea clutter. Moreover, the $\mathcal{G}_{\text{AO}}$ model has a better ability of false-alarm control than the $\mathcal{K}$ distribution, which demonstrates its usefulness and potential for the constant false alarm rate (CFAR) [36] ship detection in maritime applications [37].

In summary, both the GΓD and $\mathcal{G}_{AO}$ are promising models that show potential for describing the amplitude statistics of SAR images. However, our earlier experiments using different measured SAR data showed two significant phenomena, the first relating to the GΓD and the second to the $\mathcal{G}_{AO}$ model. We noted that the GΓD cannot match the distribution of SAR data that have a close-to-zero third-order sample log-cumulant [33], although the parameters of the GΓD can be estimated in theory, resulting in meaningless estimation results in this case. Such SAR data frequently appear for local areas of a large scene; interestingly, the $\mathcal{G}_{AO}$ model performed well in the areas where the GΓD did not work. The second point of interest is that the parameters of the $\mathcal{G}_{AO}$ model cannot be always estimated by the MoLC for all local regions of a SAR scene.

The two aforementioned phenomena imply that: (1) both the GΓD and $\mathcal{G}_{AO}$ have their inherent flaws, although the fitting performances might be adequate when they work; (2) the GΓD and $\mathcal{G}_{AO}$ models cannot be employed individually to robustly characterize the amplitude statistics of local regions in a large SAR scene with wide coverage; (3) in terms of their applicability for fitting real SAR data, the two models seem to be complementary. Note that this complementary property refers to the fact that the $\mathcal{G}_{AO}$ model performs well for those local areas where the GΓD does not work.

The goal of this section is to build a robust, reliable, and accurate scheme for statistical modeling of SAR amplitude images based on the idea that the modeling abilities of the GΓD and $\mathcal{G}_{AO}$ might be complementary.

2.2.1 $\mathcal{G}_{AO}$ Model

The PDF of the $\mathcal{G}_{AO}$ distribution with three parameters that is used for modeling SAR amplitude images is given by [35] as

$$p(x) = \frac{2^{\mu}\Gamma\left(\mu + \frac{\lambda}{2}\right)}{\beta^{-\lambda}\Gamma(\mu)\Gamma\left(\frac{\lambda}{2}\right)} \frac{x^{\mu-1}\left(\sqrt{\beta^2 + x^2} + x\right)^{1-\mu-\lambda}}{\sqrt{\beta^2 + x^2}}, \quad \mu, \beta, \lambda, x > 0 \tag{2.29}$$

where μ is a scale parameter; and β and λ are both shape parameters.

As derived in [35], the estimates of the underlying parameters in the $\mathcal{G}_{AO}$ model can be obtained by using the first three order sample log-cumulants of the $\mathcal{G}_{AO}$:

$$\ln\left(\frac{\hat{\beta}}{2}\right) - \frac{1}{2}\Psi\left(\frac{\hat{\lambda}}{2}\right) + \Psi(\hat{\mu}) - \frac{1}{2}\Psi\left(\frac{\hat{\lambda}}{2} + \hat{\mu}\right) = \hat{c}_1 \tag{2.30}$$

$$\frac{1}{4}\Psi\left(1, \frac{\hat{\lambda}}{2}\right) + \Psi(1, \hat{\mu}) - \frac{1}{4}\Psi\left(1, \frac{\hat{\lambda}}{2} + \hat{\mu}\right) = \hat{c}_2 \tag{2.31}$$

$$-\frac{1}{8}\Psi\left(2,\frac{\hat{\lambda}}{2}\right)+\Psi(2,\hat{\mu})-\frac{1}{8}\Psi\left(2,\frac{\hat{\lambda}}{2}+\hat{\mu}\right)=\hat{c}_3 \tag{2.32}$$

where $\hat{\mu}$, $\hat{\beta}$, and $\hat{\lambda}$ denote the estimates of the underling parameters μ, β, and λ, respectively.

2.2.2 *Parameter Estimates of the GΓD*

Based on the MoLC, the parameter estimators of the GΓD were expressed by [23] as

$$\frac{\Psi^3(1,\hat{\kappa})}{\Psi^2(2,\hat{\kappa})}=\frac{\hat{c}_2^3}{\hat{c}_3^2} \tag{2.33}$$

$$\hat{\nu}=\mathrm{sgn}(-\hat{c}_3)\sqrt{\Psi(1,\hat{\kappa})/\hat{c}_2} \tag{2.34}$$

$$\hat{\sigma}=\exp\{\hat{c}_1-(\Psi(\hat{\kappa})-\ln\hat{\kappa})/\hat{\nu}\} \tag{2.35}$$

where $\Psi(\cdot)$ denotes the digamma function (i.e., the logarithmic derivative of the gamma function); $\Psi(m,\cdot)$ is the mth order polygamma function (i.e., the mth order derivative of the digamma function); sgn(·) is the sign function; $\hat{\kappa}$, $\hat{\nu}$, and $\hat{\sigma}$ represent the estimates of parameters κ, ν, and σ, respectively; $\hat{c}_1$, $\hat{c}_2$, and $\hat{c}_3$ denote the first three order sample log-cumulants.

2.2.3 *Analytical Conditions of Applicability*

As the conditions of applicability of a model are related to the corresponding applicable ranges, it is desirable that the conditions can be mathematically or analytically derived for the model. In addition, the applicable conditions of a model are determined by its parameter estimation approach [34]. In this section, we derive the analytical conditions of applicability of the $\mathcal{G}_{AO}$ model and compare them with those of the GΓD.

2.2.3.1 Applicable Conditions of the GΓD

The applicable conditions of the GΓD are only related to the parameter $\hat{\kappa}$. In Eq. (2.33), because $\frac{\Psi^3(1,\hat{\kappa})}{\Psi^2(2,\hat{\kappa})}$ is continuous and monotonically increasing around $\hat{\kappa}$, it has been demonstrated that the analytical condition of applicability based on the original MoLC for the GΓD is $\hat{c}_2^3/\hat{c}_3^2 \geq 0.25$ [34].

A new estimator with respect to $\hat{\kappa}$, based on an approximation for the original MoLC estimator, was given as

$$\frac{\hat{\kappa}^2}{\hat{\kappa}+\frac{1}{2}} = \frac{\hat{c}_2^3}{\hat{c}_3^2} \tag{2.36}$$

in [33]. The new estimator, shown in Eq. (2.36), is applicable under the condition $\hat{c}_2^3/\hat{c}_3^2 < 0.25$ (refer to [33] for a detailed derivation), where the original MoLC-based estimator cannot be employed. Moreover, the fitting results for distributions of real data using the new estimator are acceptable. This means that a combination of Eqs. (2.33) and (2.36) can, in principle, solve the statistical modeling of any real data, i.e., the GΓD can theoretically be used over the entire range of $\hat{c}_2^3/\hat{c}_3^2$.

2.2.3.2 Applicable Conditions of the $\mathcal{G}_{AO}$ Model

As shown in Eqs. (2.30)–(2.32), given $\hat{\lambda}$ and $\hat{\mu}$, $\hat{\beta}$ can be obtained in terms of $\hat{c}_1$. Therefore, the applicable restrictions of the $\mathcal{G}_{AO}$ model based on the MoLC estimators are only subject to $\hat{c}_2$ and $\hat{c}_3$. Since $\Psi(1,\cdot)$ is a monotonically decreasing and positive-valued function [38], we have the following inequality:

$$\Psi\left(1, \frac{\hat{\lambda}}{2}\right) - \Psi\left(1, \frac{\hat{\lambda}}{2} + \hat{\mu}\right) > 0 \tag{2.37}$$

Thus, the first applicable condition of the $\mathcal{G}_{AO}$ model is $\hat{c}_2 > 0$ via Eq. (2.31). This condition can be satisfied by any real data, as $\hat{c}_2$ is always greater than zero. Additional restrictions to the applicability of the $\mathcal{G}_{AO}$ model should originate from the compatibility of the equations for $\hat{c}_2$ and $\hat{c}_3$, as shown in Eqs. (2.31) and (2.32).

Considering the strict monotonicity and positivity of $\Psi(1,\cdot)$ and the strict monotonicity and negativity of $\Psi(2,\cdot)$ [38], we define the inverse mapping of $\Psi(1,\cdot)$ as $\Phi_1\colon \mathbb{R}^+ \to \mathbb{R}^+$, and the inverse mapping of $\Psi(2,\cdot)$ as $\Phi_2\colon \mathbb{R}^- \to \mathbb{R}^+$. Furthermore, we state that $M = \Phi_1(\hat{c}_2)$ and $F(\cdot)$ is a function of $\hat{\lambda}$ concerning $\hat{\mu}$, i.e., $\hat{\lambda} = F(\hat{\mu})$. From Eq. (2.31), $\hat{\mu} > M$ is always true, i.e., $\hat{\mu} \in (M, +\infty)$. Substituting $\hat{\lambda} = F(\hat{\mu})$ into Eqs. (2.31) and (2.32) respectively, one obtains

$$H(\hat{\mu}) = \frac{1}{4}\Psi\left(1, \frac{F(\hat{\mu})}{2}\right) + \Psi(1,\hat{\mu}) - \frac{1}{4}\Psi\left(1, \frac{F(\hat{\mu})}{2} + \hat{\mu}\right) = \hat{c}_2 \tag{2.38}$$

$$G(\hat{\mu}) = -\frac{1}{8}\Psi\left(2, \frac{F(\hat{\mu})}{2}\right) + \Psi(2,\hat{\mu}) - \frac{1}{8}\Psi\left(2, \frac{F(\hat{\mu})}{2} + \hat{\mu}\right) = \hat{c}_3 \tag{2.39}$$

where $H(\cdot)$ is a mapping from $\hat{\mu}$ to $\hat{c}_2$, and $G(\cdot)$ is a mapping from $\hat{\mu}$ to $\hat{c}_3$. Note that $F(\hat{\mu})$ satisfies the relationships shown in Eqs. (2.38) and (2.39). However, a further analytical expression of $F(\hat{\mu})$ cannot be provided that does not influence the posterior derivations. Because $F(\cdot)$ and $G(\cdot)$ are continuously differentiable on $(M, +\infty)$, the derivatives of Eqs. (2.38) and (2.39) with respect to $\hat{\mu}$ lead to

$$\frac{1}{8}\Psi\left(2, \frac{F(\hat{\mu})}{2}\right)F'(\hat{\mu}) + \Psi(2, \hat{\mu}) - \frac{1}{4}\Psi\left(2, \frac{F(\hat{\mu})}{2} + \hat{\mu}\right)\left(\frac{F'(\hat{\mu})}{2} + 1\right) = 0 \tag{2.40}$$

$$\begin{aligned} G'(\hat{\mu}) = & -\frac{1}{16}\Psi\left(3, \frac{F(\hat{\mu})}{2}\right)F'(\hat{\mu}) + \Psi(3, \hat{\mu}) \\ & - \frac{1}{8}\Psi\left(3, \frac{F(\hat{\mu})}{2} + \hat{\mu}\right)\left(\frac{F'(\hat{\mu})}{2} + 1\right) \end{aligned} \tag{2.41}$$

From Eq. (2.40):

$$F'(\hat{\mu}) = \frac{2\Psi\left(2, \frac{F(\hat{\mu})}{2} + \hat{\mu}\right) - 8\Psi(2, \hat{\mu})}{\Psi\left(2, \frac{F(\hat{\mu})}{2}\right) - \Psi\left(2, \frac{F(\hat{\mu})}{2} + \hat{\mu}\right)} \tag{2.42}$$

Because $\Psi(2, \cdot)$ is monotonically increasing and negative-valued [38], $2\Psi\left(2, \frac{F(\hat{\mu})}{2} + \hat{\mu}\right) - 8\Psi(2, \hat{\mu}) > 0$, and $\Psi\left(2, \frac{F(\hat{\mu})}{2}\right) - \Psi\left(2, \frac{F(\hat{\mu})}{2} + \hat{\mu}\right) < 0$. Thus $F'(\hat{\mu}) < 0$, which shows that $F(\hat{\mu})$ is monotonically decreasing about $\hat{\mu}$. Next, inserting Eq. (2.42) into Eq. (2.41) results in

$$\begin{aligned} G'(\hat{\mu}) = & -\frac{1}{8}\Psi\left(3, \frac{F(\hat{\mu})}{2}\right)\left[\frac{\Psi\left(2, \frac{F(\hat{\mu})}{2} + \hat{\mu}\right) - 4\Psi(2, \hat{\mu})}{\Psi\left(2, \frac{F(\hat{\mu})}{2}\right) - \Psi\left(2, \frac{F(\hat{\mu})}{2} + \hat{\mu}\right)}\right] \\ & - \frac{1}{8}\Psi\left(3, \frac{F(\hat{\mu})}{2} + \hat{\mu}\right)\left[\frac{\Psi\left(2, \frac{F(\hat{\mu})}{2} + \hat{\mu}\right) - 4\Psi(2, \hat{\mu})}{\Psi\left(2, \frac{F(\hat{\mu})}{2}\right) - \Psi\left(2, \frac{F(\hat{\mu})}{2} + \hat{\mu}\right)}\right] \\ & + \Psi(3, \hat{\mu}) - \frac{1}{8}\Psi\left(3, \frac{F(\hat{\mu})}{2} + \hat{\mu}\right) \end{aligned} \tag{2.43}$$

Since $\Psi(2, \cdot) < 0$ and $\Psi(2, \hat{\mu}) \leq \Psi\left(2, \frac{F(\hat{\mu})}{2} + \hat{\mu}\right) < 0$, $\Psi\left(2, \frac{F(\hat{\mu})}{2} + \hat{\mu}\right) - 4\Psi(2, \hat{\mu}) > 0$ and $\Psi\left(2, \frac{F(\hat{\mu})}{2}\right) - \Psi\left(2, \frac{F(\hat{\mu})}{2} + \hat{\mu}\right) < 0$. Furthermore, because $\Psi(3, \cdot) > 0$, the following inequalities are true:

$$-\frac{1}{8}\Psi\left(3, \frac{F(\hat{\mu})}{2}\right)\left[\frac{\Psi\left(2, \frac{F(\hat{\mu})}{2} + \hat{\mu}\right) - 4\Psi(2, \hat{\mu})}{\Psi\left(2, \frac{F(\hat{\mu})}{2}\right) - \Psi\left(2, \frac{F(\hat{\mu})}{2} + \hat{\mu}\right)}\right] > 0 \tag{2.44}$$

$$-\frac{1}{8}\Psi\left(3, \frac{F(\hat{\mu})}{2} + \hat{\mu}\right)\left[\frac{\Psi\left(2, \frac{F(\hat{\mu})}{2} + \hat{\mu}\right) - 4\Psi(2, \hat{\mu})}{\Psi\left(2, \frac{F(\hat{\mu})}{2}\right) - \Psi\left(2, \frac{F(\hat{\mu})}{2} + \hat{\mu}\right)}\right] > 0 \tag{2.45}$$

As $\Psi(3, \cdot)$ is a strictly monotonically decreasing and positive-valued function [38], $\Psi(3, \hat{\mu}) - \frac{1}{8}\Psi\left(3, \frac{F(\hat{\mu})}{2} + \hat{\mu}\right) > 0$. Combining Eqs. (2.44) and (2.45) leads to $G'(\hat{\mu}) > 0$, meaning that $G(\hat{\mu})$ is monotonically increasing about $\hat{\mu}$. Since $\hat{\mu} \in (M, +\infty)$, we first consider the case of $\hat{\mu} \to \infty$. From Eq. (2.38), and based on the fact $\lim_{x\to+\infty} \Psi(n, x) = 0; \quad n = 1, 2, 3, \cdots$ [38], we obtain

$$\lim_{\hat{\mu}\to+\infty}[H(\hat{\mu})] = \frac{1}{4}\lim_{\hat{\mu}\to+\infty}\Psi\left(1, \frac{F(\hat{\mu})}{2}\right) = \frac{1}{4}\Psi\left(1, \lim_{\hat{\mu}\to+\infty}\frac{F(\hat{\mu})}{2}\right) = \hat{c}_2 \tag{2.46}$$

A further transform of Eq. (2.46) leads to

$$\lim_{\hat{\mu}\to+\infty} F(\hat{\mu}) = 2\Phi_1(4\hat{c}_2) \tag{2.47}$$

From Eq. (2.39), when $\hat{\mu} \to \infty$,

$$\begin{aligned}\lim_{\hat{\mu}\to+\infty} G(\hat{\mu}) = &-\frac{1}{8}\Psi\left(2, \lim_{\hat{\mu}\to+\infty}\frac{F(\hat{\mu})}{2}\right) + \Psi\left(2, \lim_{\hat{\mu}\to+\infty}\hat{\mu}\right)\\ &-\frac{1}{8}\Psi\left(2, \lim_{\hat{\mu}\to+\infty}\frac{F(\hat{\mu})}{2} + \lim_{\hat{\mu}\to+\infty}\hat{\mu}\right) = -\frac{1}{8}\Psi(2, \Phi_1(4\hat{c}_2))\end{aligned} \tag{2.48}$$

However, when $\mu \to M$, $\lim_{\hat{\mu}\to M} F(\hat{\mu}) = +\infty$ since $\lambda = F(\hat{\mu}) \in (0, +\infty)$ is monotonically decreasing about $\hat{\mu}$.

Therefore, from Eq. (2.39), we obtain

$$\begin{aligned}\lim_{\hat{\mu}\to M} G(\hat{\mu}) = &-\frac{1}{8}\Psi\left(2, \lim_{\hat{\mu}\to M}\frac{F(\hat{\mu})}{2}\right) + \Psi\left(2, \lim_{\hat{\mu}\to M}\hat{\mu}\right)\\ &-\frac{1}{8}\Psi\left(2, \lim_{\hat{\mu}\to M}\frac{F(\hat{\mu})}{2} + \lim_{\hat{\mu}\to M}\hat{\mu}\right) = \Psi(2, M) = \Psi(2, \Phi_1(\hat{c}_2))\end{aligned} \tag{2.49}$$

Because $G(\hat{\mu})$ is strictly monotonically increasing, from Eqs. (2.39), (2.48), and (2.49), the analytical condition of applicability of the $\mathcal{G}_{AO}$ model can finally be given as

$$\Psi(2, \Phi_1(\hat{c}_2)) < \hat{c}_3 < -\frac{1}{8}\Psi(2, \Phi_1(4\hat{c}_2)) \tag{2.50}$$

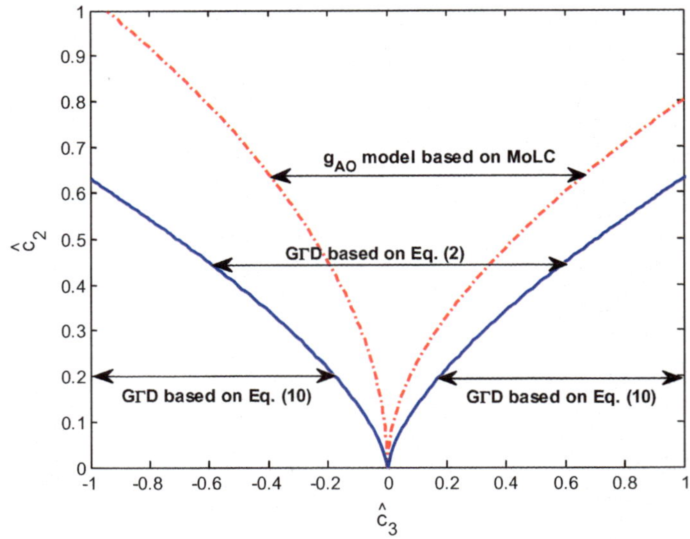

Fig. 2.7 $\hat{c}_2 \sim \hat{c}_3$ diagram indicating applicable conditions for the GΓD based on Eq. (2.33) (between the blue solid lines), the GΓD based on Eq. (2.36) (below the blue solid lines), and the $\mathcal{G}_{AO}$ model (between the red dotted lines)

As predicted by Eq. (2.50), the graph of $\hat{c}_2 \sim \hat{c}_3$ ($\hat{c}_2$ against $\hat{c}_3$) in Fig. 2.7 shows the analytical condition of applicability of the $\mathcal{G}_{AO}$ model. The applicable conditions of the GΓD based on different estimators are also plotted as comparisons in this figure. It can be observed that the applicable condition of the $\mathcal{G}_{AO}$ model is stricter than that of the GΓD. As concluded in [34], when $\hat{c}_2^3/\hat{c}_3^2 < 0.25$, the GΓD based on the estimator shown in Eq. (2.33) cannot be used, whilst an alternative estimator based on Eq. (2.36) can be employed in this case [33]. This indicates that, in theory, the applicability of the GΓD is universal for any real data because the applicable range of the GΓD can cover the entire plane of $\hat{c}_2 \sim \hat{c}_3$, as shown in Fig. 2.7.

2.2.4 *Proposed Scheme*

As discussed in the previous section, the GΓD has a wider range of applicability than the $\mathcal{G}_{AO}$ model and can, in theory, be used for any real data. However, we consistently found the GΓD to be ineffective and unable to match the distribution of real data when the value of $\hat{c}_3$ is close to zero. In this case, the data generally coincide with the range of the GΓD based on Eq. (2.33), as depicted in Fig. 2.7. We propose that the primary reason for this behavior is as follows. When $\hat{c}_3 \to 0$, $\hat{c}_2^3/\hat{c}_3^2 \to +\infty$, which leads to $\hat{\kappa} \to \infty$, as $\frac{\Psi^3(1,\hat{\kappa})}{\Psi^2(2,\hat{\kappa})}$ in Eq. (2.33) is continuous and monotonically increasing

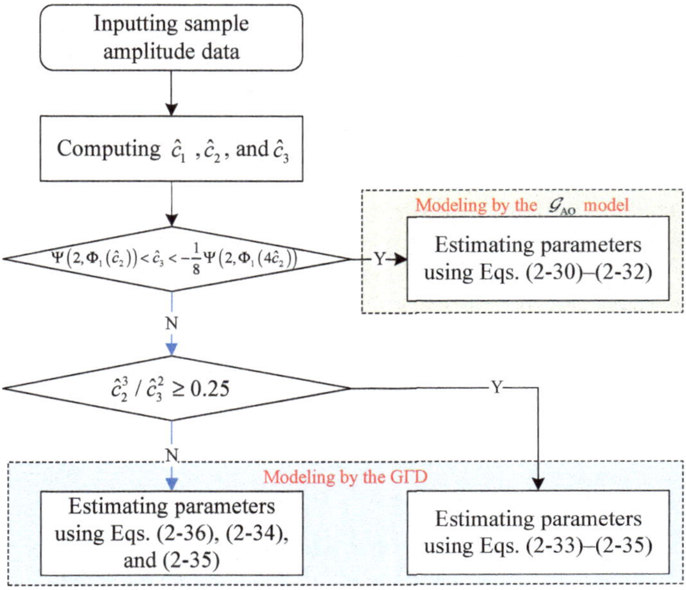

Fig. 2.8 Proposed statistical modeling scheme by combining the GΓD and $\mathcal{G}_{AO}$ model

around $\hat{\kappa}$. Therefore, the coefficient $\frac{\kappa^\kappa}{\Gamma(\kappa)}$ in the expression of the PDF of the GΓD tends to infinity and the GΓD is no longer effective. We stress that this explanation is intuitive but can be consistently observed in real data. However, a theoretical proof is difficult and not available at present.

That the GΓD does not work when the value of $\hat{c}_3$ is close to zero is an inherent property of the distribution. Thus, in this study, a statistical modeling scheme for SAR amplitude data that combines the GΓD and the $\mathcal{G}_{AO}$ model is proposed, as shown in Fig. 2.8. The scheme can be divided into the following three steps:

Step 1 We compute the sample log-cumulants $\hat{c}_1$, $\hat{c}_2$, and $\hat{c}_3$.
Step 2 If the condition that $\Psi\left(2, \Phi_1\left(\hat{c}_2\right)\right) < \hat{c}_3 < -\frac{1}{8}\Psi\left(2, \Phi_1\left(4\hat{c}_2\right)\right)$ is satisfied, the $\mathcal{G}_{AO}$ model can be used to describe the amplitude statistics of the given data and we estimate the parameters of the $\mathcal{G}_{AO}$ model according to Eqs. (2.30)–(2.32). Otherwise, we apply the GΓD as the model of the data and go to the next step.
Step 3 If $\hat{c}_2^3/\hat{c}_3^2 \geq 0.25$, the parameter estimation of the GΓD is implemented by using Eqs. (2.33)–(2.35). Otherwise, Eqs. (2.36), (2.34), and (2.35) are employed to estimate the parameters of the GΓD.

As can be observed from Figs. 2.7 and 2.8, the essence of the proposed scheme is to replace the GΓD by the $\mathcal{G}_{AO}$ model in a sub-range of applicability of the GΓD, i.e., $\Psi\left(2, \Phi_1\left(\hat{c}_2\right)\right) < \hat{c}_3 < -\frac{1}{8}\Psi\left(2, \Phi_1\left(4\hat{c}_2\right)\right)$. In this sub-range, the $\mathcal{G}_{AO}$ model is more robust than the GΓD, especially for the case where $\hat{c}_3 \to 0$. Meanwhile, the proposed scheme is universal for all real data as the applicable range of the scheme covers the entire $\hat{c}_2 \sim \hat{c}_3$ plane. In terms of the modeling ability for actual data,

[23, 31, 32], have pointed out that the GΓD exhibits better performance than many other parametric distributions in most cases and can achieve reasonable fitting results for actual distributions of land and sea clutter. Therefore, Step 3 in the proposed scheme is feasible and can guarantee a high fitting-accuracy when the condition $\Psi(2, \Phi_1(\hat{c}_2)) < \hat{c}_3 < -\frac{1}{8}\Psi(2, \Phi_1(4\hat{c}_2))$ cannot be satisfied. However, it should be noted that the operation replacing the GΓD by the $\mathcal{G}_{AO}$ model in the proposed scheme under the condition $\Psi(2, \Phi_1(\hat{c}_2)) < \hat{c}_3 < -\frac{1}{8}\Psi(2, \Phi_1(4\hat{c}_2))$ is only feasible when the following two conditions can be met. The first is that the $\mathcal{G}_{AO}$ model has a better, or at least similar, modeling ability than the GΓD when both models can be used for the real data. The second is that the $\mathcal{G}_{AO}$ model can accurately match the real data when $\hat{c}_3 \to 0$, where the GΓD does not work or the estimated parameters are meaningless. Because both the GΓD and $\mathcal{G}_{AO}$ model are empirical, and their relationship cannot be mathematically proved, the aforementioned properties will be tested in the experiments using real data in the next section.

2.2.5 *Experimental Results and Analysis*

In this section, the proposed scheme is tested on real spaceborne SAR data. The goal is to investigate how well the proposed scheme performs for fitting the distributions of both the land and sea clutter data in SAR amplitude images.

2.2.5.1 Test Data

The GΓD, the $\mathcal{G}_{AO}$ model, and the proposed scheme were tested on real SAR data from different spaceborne platforms. Two large measured scenes were selected in this experiment: one L-band Level 1.1 ALOS PALSAR scene and one C-band RADARSAT-2 SAR scene gathered from Vancouver Harbour, Canada. Originally, all image products were in SLC format and HH-polarization. Some of the key information referring to the two scenes is listed in Table 2.3.

Because the size of the original scene is too large, both these two scenes cannot be displayed in details in a limited space. Moreover, the description of SAR amplitude statistics in a homogeneous region is not a challenging task. However, the description of more heterogeneous zones, such as urban areas and regions of sea clutter in high wind conditions, is of particular interest. Thus, we randomly and manually extracted two subscenes from the two aforementioned large scenes, respectively, as shown in Figs. 2.9a and 2.10a. As a result, the subscene in Fig. 2.9a covers an extremely heterogeneous urban region, showing lots of bright scatter from buildings. The sea surfaces in Fig. 2.10a are rather rough because of a wind speed of 9–11 m/s.

Table 2.3 Scene information

Sensor	Geographic Area	Product ID	Date	Inc. Angle (°)	Resolution (m) *Slant Range* × *Azimuth*
RADARSAR-2	Vancouver Harbour	PDS_00059070	2008/04/15	19.8–21.9	5.2 × 7.6
ALOS-PALSAR	South China Sea	ALPSRP073890390	2007/06/14	38.7 (Image Center)	9.5 × 4.5

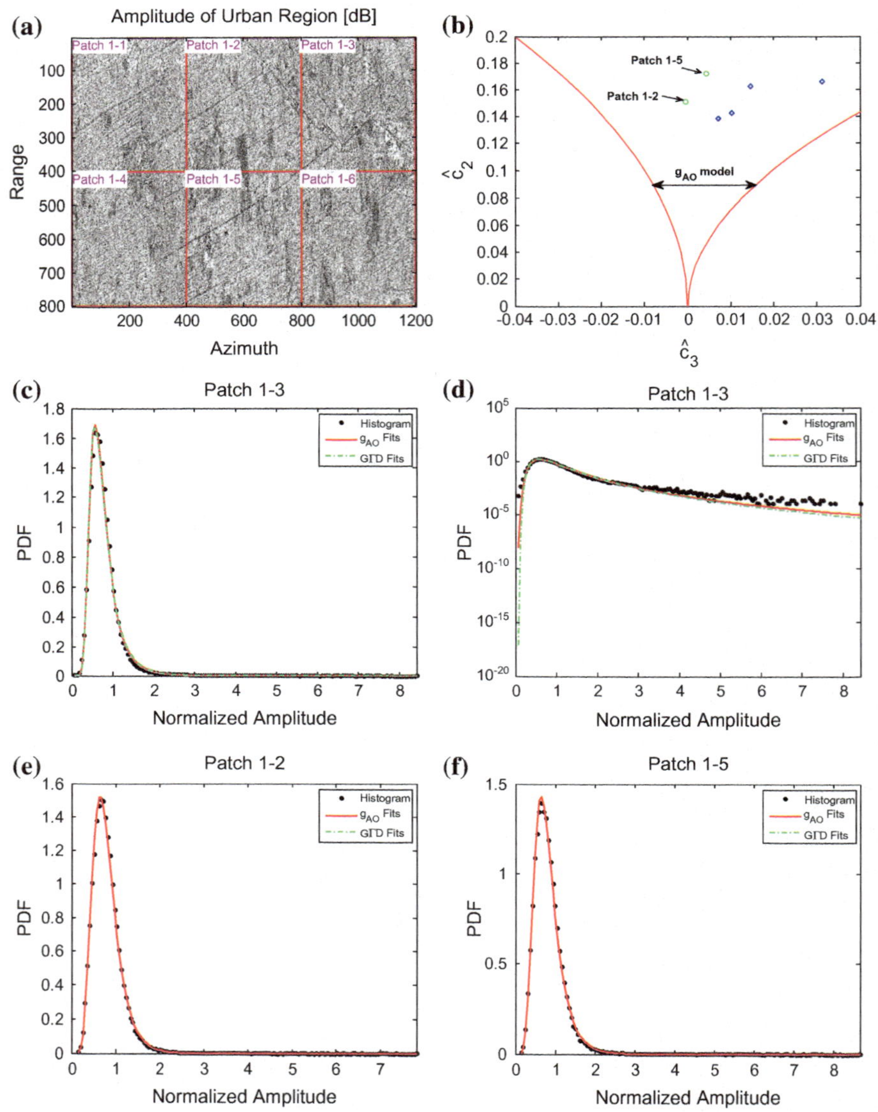

Fig. 2.9 Fitting results for urban regions of RADARSAT-2 data: **a** Subscene used in the test; **b** Locations of different patches in $\hat{c}_2 \sim \hat{c}_3$ diagram; **c** Fitting results of Patch 1-3 on a linear scale; **d** Fitting results of Patch 1-3 on a semi-log scale; **e** Fitting results of Patch 1-2; **f** Fitting results of Patch 1-5

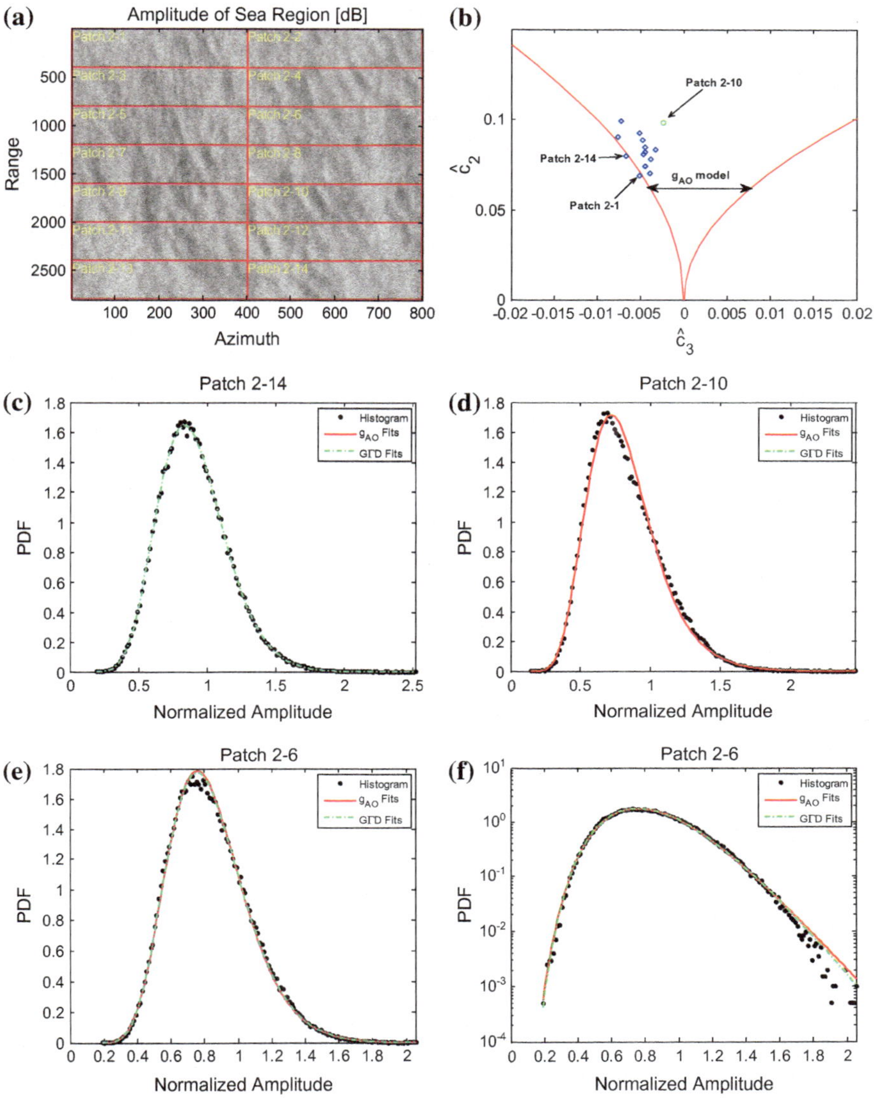

Fig. 2.10 Fitting results for sea regions of ALOS-PALSAR data: **a** Data from subscene used in the test; **b** Locations of different patches in $\hat{c}_2 \sim \hat{c}_3$ diagram; **c** Fitting results of Patch 2-14; **d** Fitting results of Patch 2-10; **e** Fitting results of Patch 2-6 on a linear scale; **f** Fitting results of Patch 2-6 on a semi-log scale

2.2.5.2 Fitting Results for Urban Regions

We first checked the effectiveness of the proposed scheme using the urban subscene [Fig. 2.9a]. The entire subscene was divided into six equally-sized blocks of 400 × 400 pixels to guarantee sufficient samples of clutter. The image blocks were numbered in turn as Patches 1-1–1-6, as shown in Fig. 2.9a. In order to allow comparison of the fitting details, the data in each patch were normalized by their mean values. The values of $\hat{c}_2$ and $\hat{c}_3$ of the six patches are listed in Table II. Based on these values, the location of each patch in the $\hat{c}_2 \sim \hat{c}_3$ diagram is plotted [Fig. 2.9b]. It can be seen that all the patches in the urban subscene fall in the applicable range of the $\mathcal{G}_{AO}$ model, and hence can be modeled using it. Specifically, the values of $\hat{c}_3$ are closer to zero in Patches 1-2 and 1-5 than in the other four patches.

Next, the GΓD and $\mathcal{G}_{AO}$ models were used to fit the amplitude data from all the patches. The results of parameter estimation in the two distributions are also provided in Table 2.4. Based on the estimates, the fitting results of several representative examples are given (the others were not provided for saving space) in Figs. 2.9c–f. Meanwhile, a quantitative comparison of fitting accuracy is also given in Table 2.4, where the commonly-used symmetric Kullback–Leibler (KL) distance [39] is employed as the goodness-of-fit measure of the global similarity between the estimated and theoretical PDFs in this study.

As the applicable range of the $\mathcal{G}_{AO}$ model is a subregion of that of the GΓD, theoretically, all patches in Fig. 3a should be able to be described by the GΓD. However, it can be clearly observed from Table 2.4 and Fig. 3 that the GΓD is not effective and cannot match the amplitude data of Patches 1-2 and 1-5. The GΓD curves are not shown in Fig. 2.9e, f because the data cannot be fitted by this model. The $\hat{c}_3$ values of these two patches are closer to zero than the values of the other four patches in this subscene (particularly, the $\hat{c}_3$ value of Patch 1-2 has the smallest distance to zero in all patches, and thus the corresponding $\hat{\kappa}$ tends to infinity) and the parameter estimation of the GΓD is meaningless, as can be seen from highlighted rows in Table 2.4. The $\mathcal{G}_{AO}$ model, on the other hand, almost perfectly matches the distributions of the amplitude statistics for Patches 1-2 and 1-5, as shown in Figs. 2.9e, f.

For Patches 1-1, 1-3, 1-4, and 1-6, both the GΓD and $\mathcal{G}_{AO}$ model perform well. The worst of these results, i.e., those with the largest KL values, are shown in Figs. 2.9c, d, where the fitting results were plotted both on linear and semi-log scales to illustrate the details of the results for fitting the whole distribution and its tails. Inspection of Table 2.4 shows that the $\mathcal{G}_{AO}$ model is slightly better at fitting the aforementioned patches than the GΓD. This can be also confirmed by the KL values between the histograms and estimated PDFs, as listed in Table 2.4, because the KL value of each patch is lower in the $\mathcal{G}_{AO}$ model than in the GΓD. It should be emphasized that both the GΓD and $\mathcal{G}_{AO}$ model perform acceptably for the four patches, as the difference of their KL values is the order of magnitude of 10^{-3} and can be neglected.

In summary, our results suggest that the $\mathcal{G}_{AO}$ model is a suitable candidate for modeling the amplitude statistics of the extremely heterogeneous urban clutter in SAR images used in this experiment. In the proposed scheme, when the real data coincide with the applicable range of the $\mathcal{G}_{AO}$ model, the $\mathcal{G}_{AO}$ model will be selected

Table 2.4 Results of parameter estimation and accuracy comparison for urban regions

Patch No.	$(\hat{c}_2, \hat{c}_3)$	GΓD $(\hat{\sigma}, \hat{\kappa}, \hat{\nu})$	$\mathcal{G}_{AO}$ $(\hat{\beta}, \hat{\lambda}, \hat{\mu})$	Proposed scheme	KL value		
					$\mathcal{G}_{AO}$	GΓD	Proposed scheme
1-1	(0.1425, 0.0102)	(0.7404, 28.2215, −0.5031)	(0.7687, 6.7364, 14.6997)	Using $\mathcal{G}_{AO}$	0.0045	0.0087	0.0045
1-2	**(0.1508, −0.0005)**	**(0.7369, 1.5053 × 10^4, 0.0210)**	**(1.0710, 8.3545, 10.3948)**	**Using $\mathcal{G}_{AO}$**	**0.0043**	/	**0.0043**
1-3	(0.1656, 0.0312)	(0.6471, 5.1263, −1.1402)	(0.4416, 4.7669, 22.1662)	Using $\mathcal{G}_{AO}$	0.0149	0.0230	0.0149
1-4	(0.1382, 0.0071)	(0.7132, 53.6209, −0.3690)	(0.8037, 7.3321, 13.9101)	Using $\mathcal{G}_{AO}$	0.0040	0.0082	0.0040
1-5	**(0.1716, 0.0043)**	**(0.7461, 275.7434, −0.1455)**	**(0.9720, 6.7703, 9.9956)**	**Using $\mathcal{G}_{AO}$**	**0.0047**	/	**0.0047**
1-6	(0.1622, 0.0147)	(0.6908, 20.1435, −0.5602)	(0.6956, 5.8856, 13.5048)	Using $\mathcal{G}_{AO}$	0.0042	0.0081	0.0042

automatically, as summarized in Table 2.4. In this experiment, the fitting results of the $\mathcal{G}_{\mathrm{AO}}$ model are just those of the proposed scheme, which means that the same KL values can be obtained by the proposed scheme with by the $\mathcal{G}_{\mathrm{AO}}$ model, see Table 2.4. In summary, the proposed scheme can not only deal with patches where the GΓD cannot be used (i.e., Patches1-2 and 1-5), but also provide a slightly better fitting performance than the GΓD in terms of its modeling abilities. Thus, the experiment using the RADARSAT-2 data validates the feasibility and effectiveness of the proposed scheme.

2.2.5.3 Fitting Results for Sea Regions

In the second experiment, we investigate the effectiveness of the proposed scheme for the sea clutter. A rough sea region was divided into 14 equally-sized image blocks of 400 × 400 pixels, which were numbered Patches 2-1–2-14, as shown in Fig. 2.10a. Figure 4b shows the locations of different patches in the $\hat{c}_2 \sim \hat{c}_3$ diagram. Table 2.5 lists the values of $\hat{c}_2$, $\hat{c}_3$, $\Psi(2, \Phi_1(\hat{c}_2))$, $-\frac{1}{8}\Psi(2, \Phi_1(4\hat{c}_2))$, the results of parameter estimation by the GΓD and $\mathcal{G}_{\mathrm{AO}}$ model, as well as the KL values of the two distributions and the proposed scheme for fitting all patches in this experiment.

As can be seen from highlighted rows in Table 2.5, only two patches, Patches 2-1 and 2-14, cannot satisfy the condition that $\Psi(2, \Phi_1(\hat{c}_2)) < \hat{c}_3 < -\frac{1}{8}\Psi(2, \Phi_1(4\hat{c}_2))$. This is also shown by Fig. 2.10b, where these two patches are located outside the applicable range of the $\mathcal{G}_{\mathrm{AO}}$ model, which confirms the correctness of theoretical derivations. In other words, Patches 2-1 and 2-14 cannot be modeled by the $\mathcal{G}_{\mathrm{AO}}$ model, which is consistent with the fitting results shown in Fig. 2.10c and Table 2.5. Note that the fitting results of the $\mathcal{G}_{\mathrm{AO}}$ model are not shown in Fig. 2.10c because the model's parameters cannot be estimated. In addition, the fitting results of Patch 2-1 do not be provided for saving space.

From Fig. 2.10b and Table 2.5, it can also be observed that the $\hat{c}_3$ value of Patch 2-10 is the closest to zero of all the patches investigated in this experiment, leading to a large estimated value of $\hat{\kappa}$ for this patch in the GΓD, as highlighted in Table 2.5. Similarly, although the parameters of the GΓD can be estimated in principle, the estimates are meaningless because the GΓD cannot match the distribution of the real data, and hence the fitting result of the GΓD is omitted from Fig. 2.10d. However, the $\mathcal{G}_{\mathrm{AO}}$ model can provide a good match in this situation, as illustrated by Fig. 2.10d.

For the patches for which both the GΓD and $\mathcal{G}_{\mathrm{AO}}$ model can be employed, the two distributions exhibit a similar modeling ability and both perform as predicted. This can be seen intuitively based on a visual inspection of even the worst fitting results with largest KL values (i.e., Patch 2-6), which is shown in Figs. 2.10e, f. Overall, in contrast to the first experiment, the GΓD exhibits a slightly better fitting accuracy than the $\mathcal{G}_{\mathrm{AO}}$ model for patches where both the distributions can be used. However, we stress that both the GΓD and $\mathcal{G}_{\mathrm{AO}}$ model perform acceptably for all such patches, as the difference between their KL values is small enough to be negligible. This confirms that the GΓD can be indeed replaced by the $\mathcal{G}_{\mathrm{AO}}$ model when both are effective for real data.

Table 2.5 Results of parameter estimation and accuracy comparison for sea regions

Patch No.	$\left(\hat{c}_2, \hat{c}_3, \Psi(2, \Phi_1(\hat{c}_2)), -\frac{1}{8}\Psi(2, \Phi_1(4\hat{c}_2))\right)$	GΓD $(\hat{\sigma}, \hat{\kappa}, \hat{\nu})$	$\mathcal{G}_{AO}$ $\left(\hat{\beta}, \hat{\lambda}, \hat{\mu}\right)$	Proposed scheme	KL value		
					$\mathcal{G}_{AO}$	GΓD	Proposed scheme
2-1	(0.0690, −0.0051, −0.0048, 0.0095)	(0.9194, 13.2317, 1.0667)	/	Using GΓD	/	0.0015	0.0015
2-2	(0.0702, −0.0039, −0.0049, 0.0098)	(0.8816, 23.2336, 0.7915)	(3.1736, 44.0158, 15.8126)	Using $\mathcal{G}_{AO}$	0.0027	0.0023	0.0027
2-3	(0.0745, −0.0045, −0.0056, 0.0110)	(0.8583, 21.1417, 0.8062)	(3.2090, 43.3747, 14.8489)	Using $\mathcal{G}_{AO}$	0.0022	0.0018	0.0022
2-4	(0.0809, −0.0047, −0.0065, 0.0130)	(0.8474, 24.4821, 0.7178)	(2.5875, 32.0165, 14.1491)	Using $\mathcal{G}_{AO}$	0.0031	0.0024	0.0031
2-5	(0.0782, −0.0039, −0.0061, 0.0121)	(0.8147, 32.3980, 0.6333)	(2.1549, 28.2396, 15.0889)	Using $\mathcal{G}_{AO}$	0.0024	0.0018	0.0024
2-6	(0.0823, −0.0045, −0.0068, 0.0134)	(0.8225, 27.8858, 0.6662)	(2.2904, 28.4659, 14.1873)	Using $\mathcal{G}_{AO}$	0.0065	0.0053	0.0065
2-7	(0.0927, −0.0051, −0.0086, 0.0170)	(0.7914, 31.4070, 0.5909)	(1.9644, 22.4477, 13.0111)	Using $\mathcal{G}_{AO}$	0.0055	0.0040	0.0055
2-8	(0.0837, −0.0033, −0.0070, 0.0139)	(0.8393, 55.4730, 0.4663)	(1.7895, 20.9938, 15.0629)	Using $\mathcal{G}_{AO}$	0.0054	0.0040	0.0054
2-9	(0.0995, −0.0073, −0.0099, 0.0196)	(0.8325, 19.2019, 0.7330)	(2.6094, 27.1359, 11.5365)	Using $\mathcal{G}_{AO}$	0.0052	0.0040	0.0052

(continued)

Table 2.5 (continued)

Patch No.	$\left(\hat{c}_2, \hat{c}_3, \Psi(2, \Phi_1(\hat{c}_2)), -\frac{1}{8}\Psi(2, \Phi_1(4\hat{c}_2))\right)$	GΓD $(\hat{\sigma}, \hat{\kappa}, \hat{\nu})$	$\mathcal{G}_{AO}$ $\left(\hat{\beta}, \hat{\lambda}, \hat{\mu}\right)$	Proposed scheme	KL value		
					$\mathcal{G}_{AO}$	GΓD	Proposed scheme
2-10	**(0.0983, −0.0024, −0.0097, 0.0191)**	**(0.7880, 169.3273, 0.2455)**	**(1.3584, 14.5217, 14.1269)**	**Using $\mathcal{G}_{AO}$**	**0.0124**	/	**0.0124**
2-11	(0.0904, −0.0076, −0.0082, 0.0162)	(0.8973, 13.2086, 0.9328)	(5.7763, 66.0018, 11.8121)	Using $\mathcal{G}_{AO}$	0.0025	0.0023	0.0025
2-12	(0.0890, −0.0047, −0.0079, 0.0157)	(0.8701, 32.3486, 0.5941)	(2.1704, 23.4463, 13.5147)	Using $\mathcal{G}_{AO}$	0.0045	0.0032	0.0045
2-13	(0.0847, −0.0045, −0.0072, 0.0142)	(0.9028, 30.6975, 0.6252)	(2.3563, 25.7886, 14.0006)	Using $\mathcal{G}_{AO}$	0.0044	0.0032	0.0044
2-14	(0.0801, −0.0067, −0.0064, 0.0127)	(0.9053, 11.9685, 1.0429)	/	Using GΓD	/	0.0012	0.0012

The model to be implemented by our proposed scheme is summarized in Table 2.5. In the case where the GΓD cannot be used, e.g., for Patch 2-10, the proposed scheme automatically selects the $\mathcal{G}_{\text{AO}}$ model, which can provide an acceptably good fit. In the case that the $\mathcal{G}_{\text{AO}}$ model cannot be estimated, e.g., for Patches 2-1 and 2-14, the proposed scheme automatically selects the GΓD, which has a high fitting-accuracy for real data. In the case that both the GΓD and $\mathcal{G}_{\text{AO}}$ models can be applied, the proposed scheme automatically selects the $\mathcal{G}_{\text{AO}}$ model. As the GΓD and $\mathcal{G}_{\text{AO}}$ models show very similar performance on real data in this case, the fitting accuracy of the proposed scheme can be assured. Therefore, in summary, the proposed scheme is robust, reliable, and accurate for fitting all data in this experiment. This is a distinct improvement over the GΓD and $\mathcal{G}_{\text{AO}}$ models alone because, as we have demonstrated in this study, neither of the distributions can be employed individually to robustly characterize the amplitude statistics of local regions in a large SAR scene.

2.3 An Improved Scheme for Parameter Estimation of G^0 Distribution Model in High-Resolution SAR Images

The K and G^0 distributions are just the two special forms of the G distribution [16], where the K distribution is suitable for modeling heterogeneous regions while the G^0 distribution is suitable for modeling multilook clutter with widely varying degrees of homogeneity. Compared with the G distribution, the G^0 distribution does not involve complex Bessel functions. So the parameter estimation of the G^0 distribution is relatively easy, and the computational complexity is low. Consequently, the G^0 distribution has become one of the most promising statistical models in recent years [29, 30].

The essential part of the G^0 distribution is parameter estimation. Several strategies have been proposed to deal with parameter estimation problem, such as the method of moments (MoM) [16, 29, 31–33] and mellin transform (MT) based parameter estimation [34]. However, the fitting precision of the MoM is low. The MT based method can not model extremely homogeneous regions although it has a high estimation precision.

2.3.1 *The* $\mathbf{G^0}$ *Model*

In order to retrieve the "unspeckled" radar backscatter component from an observed SAR image data, a model combining an underlying RCS component with an uncorrelated speckle component is used. The most commonly used model is the product model, which expresses the observed component Z as the product of the RCS component X and the speckle noise component Y

$$Z = X \cdot Y \tag{2.51}$$

Frery et al. [16] deduced the G model by assuming a generalized inverse Gaussian distribution for the RCS component and a Gamma distribution for the speckle component of multi-look SAR images. The intensity and amplitude PDF of the G distribution are

$$\begin{aligned} p_{Z_I}(x) &\sim G_I(\alpha, \gamma, \lambda, n) \\ &= \frac{n^n(\lambda/\gamma)^{\alpha/2}}{\Gamma(n)K_\alpha(2\sqrt{\lambda\gamma})} x^{n-1}(\frac{\gamma + nx}{\lambda})^{(\alpha-n)/2} K_{\alpha-n}(2\sqrt{\lambda(\gamma + nx)}) \end{aligned} \tag{2.52}$$

$$\begin{aligned} p_{Z_A}(x) &\sim G_A(\alpha, \gamma, \lambda, n) \\ &= \frac{2n^n(\lambda/\gamma)^{\alpha/2}}{\Gamma(n)K_\alpha(2\sqrt{\lambda\gamma})} x^{2n-1}(\frac{\gamma + nx^2}{\lambda})^{(\alpha-n)/2} K_{\alpha-n}(2\sqrt{\lambda(\gamma + nx^2)}) \end{aligned} \tag{2.53}$$

where $p_{Z_I}(x)$ is the intensity PDF, and $p_{Z_A}(x)$ is the amplitude PDF, $x \in \mathbb{R}$ is the RCS component, and n is the equivalent number of looks, and λ, γ are the scale parameters which have a relationship with the mean energy of the observed region, and K_α denotes the modified Bessel function of the third kind, and α indicates the degree of heterogeneity [16], and $\Gamma(\cdot)$ denotes the Gamma function.

The K distribution is a particular form of the G model, which assumes that RCS component and the speckle component both as Gamma distribution. If $\gamma \to 0$, the intensity and amplitude PDF of K distribution can be deduced from Eqs. (2.51) and (2.52) respectively

$$p_{Z_I}(x) \sim K_I(\alpha, \lambda, n) = \frac{2\lambda n}{\Gamma(\alpha)\Gamma(n)}(\lambda nx)^{(\alpha+n)/2-1} K_{\alpha-n}(2\sqrt{\lambda nx}), \ \alpha, \lambda, n, x > 0 \tag{2.54}$$

$$p_{Z_A}(x) \sim K_A(\alpha, \lambda, n) = \frac{4\lambda nx}{\Gamma(\alpha)\Gamma(n)}(\lambda nx^2)^{(\alpha+n)/2-1} K_{\alpha-n}(2x\sqrt{\lambda n}), \ \alpha, \lambda, n, x > 0 \tag{2.55}$$

The Gamma distribution can well describe the characteristics of the RCS fluctuations of heterogeneous terrains in high resolution SAR images. But it can't model extremely heterogeneous terrains like urban regions containing many buildings.

The G^0 distribution is another particular form of the G model, which assumes that the RCS component as SRGIG distribution and the speckle component as Gamma distribution. If $\lambda \to 0$, the intensity and amplitude PDF of the G^0 distribution can be deduced from Eqs. (2.51) and (2.52) respectively

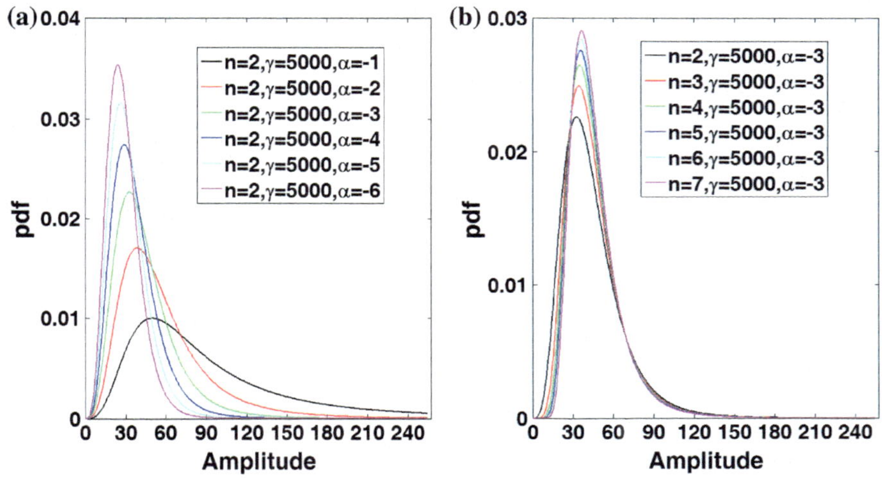

Fig. 2.11 The amplitude PDF of G^0 distribution with $\gamma = 5000$ for different values of n and α. **a** $n = 2$, $\alpha = \{-1, -2, -3, -4, -5, -6\}$. **b** $n = \{2, 3, 4, 5, 6, 7\}$, $\alpha = -3$

$$p_{Z_I}(x) \sim G_I^0(\alpha, \gamma, n) = \frac{n^n \Gamma(n-\alpha) x^{n-1}}{\gamma^\alpha \Gamma(n)\Gamma(-\alpha)(\gamma + nx)^{n-\alpha}}, \quad -\alpha, \gamma, n, x > 0 \tag{2.56}$$

$$p_{Z_A}(x) \sim G_A^0(\alpha, \gamma, n) = \frac{2n^n \Gamma(n-\alpha) x^{2n-1}}{\gamma^\alpha \Gamma(n)\Gamma(-\alpha)(\gamma + nx^2)^{n-\alpha}}, \quad -\alpha, \gamma, n, x > 0 \tag{2.57}$$

The SRGIG distribution is appropriate for modeling both heterogeneous and extremely heterogeneous regions. Moreover, the PDF given in Eqs. (2.56) and (2.57) don't involve the complex Bessel functions. Thus, it potentially means simple parameter estimation and is regarded as the universal model for speckled imagery. Figure 2.11a shows some examples of Eq. (2.57) with $n = 2$ for different values of α, and Fig. 2.11b shows some examples of Eq. (2.57) with $\alpha = -3$ for different values of n. It can be seen that parameter α is able to provide more flexibility to control the model shape than parameter n.

As mentioned in the introduction section, the existing efficient parameter estimation techniques for G^0 distribution are the MoM and MT based methods. In order to simplify the introduction, we only discuss parameter estimation for the intensity PDF of the G^0 distribution because the estimated values are same both for intensity and amplitude data.

2.3.2 MoM Based Parameter Estimation

The kth order moments of the G^0 distribution are given by

$$E(x^k) = \int_0^{+\infty} x^k G_I^0(\alpha, \gamma, n) dx$$
$$= \frac{n^n \Gamma(n-\alpha)}{\gamma^\alpha \Gamma(n)\Gamma(-\alpha)} \int_0^{+\infty} \frac{x^{n+k-1}}{(\gamma + nx)^{n-\alpha}} dx, \quad -\alpha, \gamma, n, x > 0 \tag{2.58}$$

After simplification, Eq. (2.56) can be expressed as

$$E(x^k) = \left(\frac{\gamma}{n}\right)^k \frac{\Gamma(-\alpha-k)\Gamma(n+k)}{\Gamma(n)\Gamma(-\alpha)}, \quad -\alpha, \gamma, n, x > 0 \tag{2.59}$$

In order to estimate the parameters α, γ and n, two well-known moment estimation methods exist [16, 29]. For intensity data, and assuming the equivalent number of looks n known, and using $k = 1/2, 1$ [16]. The estimator for the parameters of the G^0 distribution are given

$$\begin{cases} \frac{\Gamma^2(-\alpha-1/2)\Gamma^2(n+1/2)}{\Gamma(-\alpha-1)\Gamma(n+1)\Gamma(-\alpha)\Gamma(n)} = \frac{E^2(x^{1/2})}{E(x)} \\ \gamma = n\left(\frac{E(x)\Gamma(-\alpha)\Gamma(n)}{\Gamma(-\alpha-1)\Gamma(n+1)}\right)^2 \end{cases}, \quad -\alpha, \gamma, n, x > 0 \tag{2.60}$$

We call this moment estimation method as the classical MoM.

Assuming the equivalent number of looks n known, and using $k = 1, 2$ [29]. The estimator for the parameters of the G^0 distribution are given

$$\begin{cases} E(x) = \frac{\gamma}{-(\alpha+1)} \\ E(x^2) = \frac{\gamma^2(n+1)}{n\alpha(\alpha+1)} \end{cases}, \quad \alpha < -1, \gamma, n, x > 0 \tag{2.61}$$

Both of these two moment estimation methods are taken the equivalent number of looks n as known. However, the equivalent number of looks of different regions varies greatly in a whole SAR image. It is reasonless to replace the mean number of looks as the number of looks of different regions, and is necessary to take n as an estimation parameter.

In order to estimate three parameters α, γ and n, three equations are needed at least. Based on the property $\Gamma(x+1) = x\Gamma(x)$ of the gamma function $\Gamma(x) = \int_0^\infty t^{x-1} e^{-t} dt \ (x > 0)$, we deduce the iterative form of moment calculation as

$$E(x^{k+1}) = -\frac{\gamma}{n} \frac{n+k}{\alpha+k+1} E(x^k), \quad -\alpha, \gamma, n, x > 0 \tag{2.62}$$

Based on Eq. (2.62) and by using k = 1, 2 and 3, the estimator for the parameters of the G^0 distribution are given

$$\begin{cases} E(x) = -\frac{\gamma}{\alpha+1} \\ E(x^2) = \frac{(n+1)(a+1)}{n(a+2)} E^2(x) \\ E(x^3) = \frac{(n+2)(a+1)}{n(a+3)} E^2(x)E(x) \end{cases}, \quad -\alpha, \gamma, n, x > 0 \tag{2.63}$$

We call this moment estimation method as the extended MoM.

However, these two MoM based parameter estimation methods of the G^0 distribution mentioned above may be not sufficient to model regions in high-resolution SAR images.

2.3.3 MT Based Parameter Estimation

All the principal statistical moments of MT can be deduced from results of fourier transform [35]. The main functions of the G^0 distribution like First second-kind characteristic function, Second second-kind characteristic function, rth order second-kind characteristic moment and rth order second-kind characteristic cumulant are as follows respectively

$$\begin{aligned} \phi_I(x) &= \mathrm{MT}[p_{Z_I}](x) = \int_0^{+\infty} u^{x-1} p_{Z_I}(u)du = \int_0^{+\infty} u^{x-1} G_I^0(\alpha, \gamma, n)du \\ &= (\frac{\gamma}{n})^{x-1} \frac{\Gamma(n+x-1)\Gamma(-\alpha-x+1)}{\Gamma(n)\Gamma(-\alpha)}, \ \alpha < 1-x, \gamma, n, x > 0 \end{aligned} \tag{2.64}$$

$$\varphi_I(x) = \log(\phi_I(x)) \tag{2.65}$$

$$\widetilde{m_r} = \frac{d^r \phi_I(x)}{dx^r}|_{x=1} = \int_0^{+\infty} (\log u)^r p_{Z_I}(u)du \tag{2.66}$$

$$\widetilde{k_r} = \frac{d^r \varphi_I(x)}{dx^r}|_{x=1} \tag{2.67}$$

Combining Eqs. (2.64) and (2.65), the first 3th order second-kind characteristic moment can be deduced

$$\begin{cases} \widetilde{m_1} = \log(\gamma/n) + \Psi(n) - \Psi(-\alpha) \\ \widetilde{m_2} = \Psi^{(1)}(n) + \Psi^{(1)}(-\alpha) + \widetilde{m_1}^2 \\ \widetilde{m_3} = \Psi^{(2)}(n) - \Psi^{(2)}(-\alpha) + 3(\Psi^{(1)}(n) + \Psi^{(1)}(-\alpha))\widetilde{m_1} + \widetilde{m_1}^2 \end{cases} \tag{2.68}$$

where $\Psi(x) = \Gamma'(x)/\Gamma(x)$ $(x > 0)$ is the digamma function, and $\Psi^{(k)}(x) = (-1)^{k+1}\int_0^\infty \frac{t^k e^{-xt}}{1-e^{-t}}dt$ is the kth order digamma function which is also called the polygamma function. As for regular moments, the relationships between the first 3th order second-kind characteristic moment and characteristic cumulant are [35]

$$\begin{cases} \widetilde{k_1} = \widetilde{m_1} \\ \widetilde{k_2} = \widetilde{m_2} - \widetilde{m_1}^2 \\ \widetilde{k_3} = \widetilde{m_3} - 3\widetilde{m_1}\widetilde{m_2} + 2\widetilde{m_1}^3 \end{cases} \tag{2.69}$$

Combining Eqs. (2.68) and (2.69), the following equations are obtained

$$\begin{cases} \widetilde{k_1} = \log(\gamma/n) + \Psi(n) - \Psi(-\alpha) \\ \widetilde{k_2} = \Psi^{(1)}(n) + \Psi^{(1)}(-\alpha) \\ \widetilde{k_3} = \Psi^{(2)}(n) - \Psi^{(2)}(-\alpha) \end{cases} \tag{2.70}$$

Define $x_1, x_2, \cdots, x_N$ as N samples of the observed data, the first 3th order second-kind characteristic cumulant are actually calculated as

$$\begin{cases} \widetilde{k_1} = \frac{1}{N}\sum_{i=1}^{N} \log(x_i) \\ \widetilde{k_2} = \frac{1}{N}\sum_{i=1}^{N} [(\log(x_i) - \widetilde{k_1})^2] \\ \widetilde{k_3} = \frac{1}{N}\sum_{i=1}^{N} [(\log(x_i) - \widetilde{k_1})^3] \end{cases} \tag{2.71}$$

The solution of Eq. (2.70) is usually done by calculating the second and third expression firstly. Define two new functions by replacing α as x in Eq. (2.70)

$$\begin{cases} f_1(x) = (\Psi^{(1)})^{-1}(\widetilde{k_2} - \Psi^{(1)}(-x)) \\ f_2(x) = (\Psi^{(2)})^{-1}(\widetilde{k_3} + \Psi^{(2)}(-x)) \end{cases}, \quad x < 0 \tag{2.72}$$

where $(\Psi^{(1)})^{-1}(x)$ is the inverse 1th order digamma function, and $(\Psi^{(2)})^{-1}(x)$ is the 2th order digamma function. According to the property of the polygamma function, $\Psi^{(1)}(x)$ is completely monotonic increase while $\Psi^{(2)}(x)$ is completely monotonic decrease. Based on this fact, it is easy to see that $f_1(x)$ is completely monotonic decrease while $f_2(x)$ is completely monotonic increase. The intersection point of $f_1(x)$ and $f_2(x)$ is the estimated value of α. After calculating the parameter α, n and γ are obtained by Eq. (2.70).

The MT based parameter estimation method can model homogeneous, heterogeneous and extremely heterogeneous regions of SAR image, and has a high estimation precision. The key point of all the merits mentioned above is based on the fact that functions $f_1(x)$ and $f_2(x)$ have an intersection point. However, the parameter α is a large number when the observed region is extremely homogeneous. If $|\alpha| \to \infty$, $\Psi^{(1)}(x) \to 0$ and $\Psi^{(2)}(x) \to 0$, consequently $f_1(\infty) \to (\Psi^{(1)})^{-1}(\widetilde{k_2})$ and $f_2(\infty) \to (\Psi^{(1)})^{-1}(\widetilde{k_3})$ would be constant. If $f_1(\infty) > f_2(\infty)$, there wouldn't exist an intersection point, and Eq. (2.70) is unsolved.

2.3.4 *Our Proposed Parameter Estimation*

In order to obtain the parameters estimation of the G^0 distribution and satisfy the constraints of these parameters, we propose a novel parameter estimation method, which combines the merits both from the MoM and MT parameter estimation. Firstly, we convert Eq. (2.61) into

$$\begin{cases} \alpha = \frac{2nE(x^2)-(n+1)E^2(x)}{(n+1)E^2(x)-nE(x^2)} \\ \gamma = -(\alpha+1)E(x) \end{cases}, \quad \alpha < -1, \gamma, n, x > 0 \tag{2.73}$$

Then, a new equation is formed by substituting the Eq. (2.73) for the second and third expressions of Eq. (2.70)

$$\begin{cases} \log(\gamma/n) + \Psi(n) - \Psi(-\alpha) = \widetilde{k_1} \\ \alpha = \frac{2nE(x^2)-(n+1)E^2(x)}{(n+1)E^2(x)-nE(x^2)} \\ \gamma = -(\alpha+1)E(x) \end{cases}, \quad \alpha < -1, \gamma, n, x > 0 \tag{2.74}$$

Also, we can deduce from Eq. (2.73) that

$$\frac{\gamma}{n} = \frac{E(x^2)E(x)}{nE(x^2) - (n+1)E^2(x)}, \gamma, n, x > 0 \tag{2.75}$$

Thirdly, the parameter n is obtained by taking Eq. (2.75) and the second expression of Eq. (2.74) into the fist expression of Eq. (2.74).

Finally, the parameter α and γ are obtained subsequently by solving Eq. (2.74).

Our parameter estimation method of the G^0 distribution only requires calculating the 1th order and 2th order moments while the MoM based parameter estimation method needs to the 3th order moment. Additionally, our method avoids the instance of no solution when modeling the extremely homogeneous regions in high resolution SAR images.

For the convenience of comparison, the parameter estimation methods of the G^0 distribution mentioned above are summarized in Table 2.6.

Table 2.6 The expressions of four methods for parameter estimation of the G^0 distribution mentioned in this chapter

Methods	Expressions
The classical MoM	$k=1/2,1\begin{cases}\frac{\Gamma^2(-\alpha-1/2)\Gamma^2(n+1/2)}{\Gamma(-\alpha-1)\Gamma(n+1)\Gamma(-\alpha)\Gamma(n)}=\frac{E^2(x^{1/2})}{E(x)}\\ \gamma=n(\frac{E(x)\Gamma(-\alpha)\Gamma(n)}{\Gamma(-\alpha-1)\Gamma(n+1)})^2\end{cases}, \quad -\alpha,\gamma,n,x>0$
The extended MoM	$k=1,2,3\begin{cases}E(x)=-\frac{\gamma}{\alpha+1}\\ E(x^2)=\frac{(n+1)(\alpha+1)}{n(\alpha+2)}E^2(x)\\ E(x^3)=\frac{(n+2)(\alpha+1)}{n(\alpha+3)}E^2(x)E(x)\end{cases}, \quad -\alpha,\gamma,n,x>0$
MT	$\begin{cases}\log(\frac{\gamma}{n})+\Psi(n)-\Psi(-\alpha)=\frac{1}{N}\sum_{i=1}^{N}\log(x_i)=\widetilde{k_1}\\ \Psi^{(1)}(n)+\Psi^{(1)}(-\alpha)=\frac{1}{N}\sum_{i=1}^{N}[(\log(x_i)-\widetilde{k_1})^2]\\ \Psi^{(2)}(n)-\Psi^{(2)}(-\alpha)=\frac{1}{N}\sum_{i=1}^{N}[(\log(x_i)-\widetilde{k_1})^3]\end{cases}$
Our proposed	$\begin{cases}\log(\gamma/n)+\Psi(n)-\Psi(-\alpha)=\widetilde{k_1}\\ \alpha=\frac{2nE(x^2)-(n+1)E^2(x)}{(n+1)E^2(x)-nE(x^2)}\\ \gamma=-(\alpha+1)E(x)\end{cases}, \quad \alpha<-1,\gamma,n,x>0$

2.3.5 *Results and Discussion*

The proposed parameter estimation method is test in real SAR images. Two SAR images taken from different sensors are used in the experiment. The first one was collected by the Ku-band miniSAR of Sandia National Laboratories over the region of Eubank Gated Entrance, KAFB, USA on 19 May, 2005. Its size is 2510 × 1638 pixels with a pixel size of about 0.1 m (see Fig. 2.12). The other was taken by an airborne X-band SAR of China Electronics Technology Group Corporation No. 38 Research Institute over the region of Hefei, Anhui province, China on 6 and 8 November, 2005. Its size is 1840 × 907 pixels with a pixel size of about 1 m (see Fig. 2.13). Four representative regions (see R1 to R4 in Figs. 2.11 and 2.12 respectively) are chosen from each SAR image, where R1 is the extremely homogeneous region, and R2 is the homogeneous region, and R3 is the heterogeneous region, and R4 is the extremely heterogeneous region. All the experiments are run on P(R) dual-core 2.8 GHz CPU, with 2 GB SDRAM, and the software platform is Matlab R2007b.

2.3.5.1 Parameter Estimation Experiments

In order to test the efficiency of our proposed method, the comparison with the other parameter estimation methods is made like the classical MoM [29], our deduced the extended MoM and MT [34]. Table 2.7 shows the results of these parameter

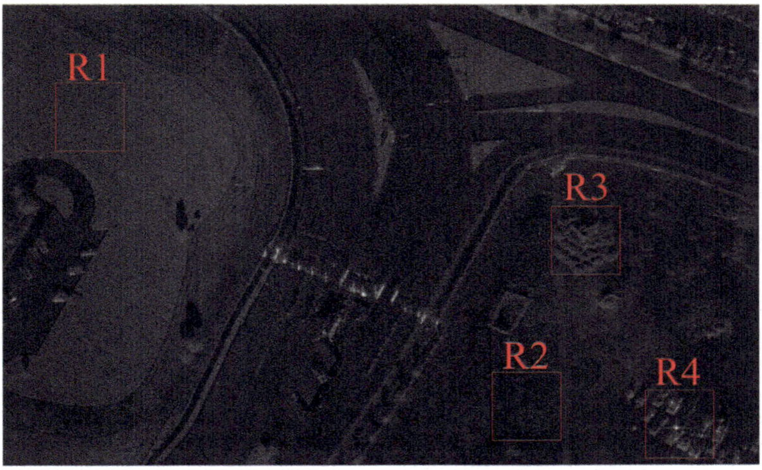

Fig. 2.12 Original SAR image collected by the miniSAR of Sandia National Laboratories, and red squares (R1 to R4) indicate the four selected representative regions, which are ground, vegetation, trees and buildings respectively

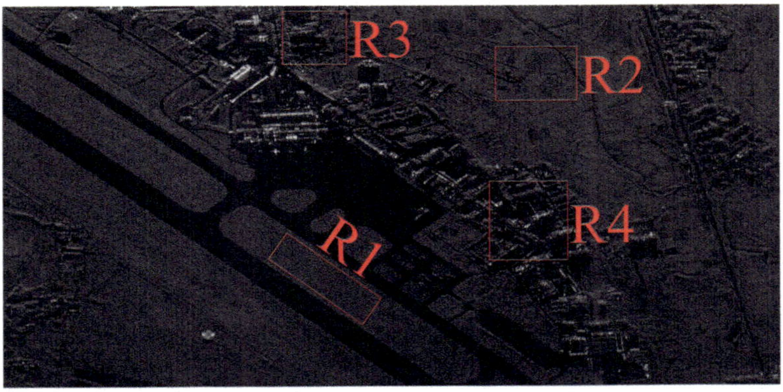

Fig. 2.13 Original SAR image collected by an airborne SAR of China Electronics Technology Group Corporation No. 38 Research Institute, and red squares (R1 to R4) indicate the four selected representative regions, which are airdrome ground, vegetation, trees and buildings respectively

estimation algorithms of four representative regions in Fig. 2.12 and Table 2.8 shows the results of these parameter estimation algorithms of four representative regions in Fig. 2.13.

According to Eq. (2.60), the estimated parameters must meet the requirement that α is negative and γ, n, x are positive. However, it can be seen that the parameters like α and γ obtained by the classical MoM exceed the range of the requirement of Eq. (2.60) both in Tables 2.7 and 2.8. The parameters like α and γ obtained by the extended MoM seldom exceed the range of definition except R1 in Table 2.7.

Table 2.7 Parameter estimation results in four regions of Fig. 2.12

Regions	Methods	Parameters		
		n	α	γ
R1	The classical MoM	4.4	2.9×10^5	-6.8×10^8
	The extended MoM	4.4	1.7×10^3	-4.0×10^6
	MT	Unsolved	Unsolved	Unsolved
	Our proposed	4.5	−577.5	1.3×10^6
R2	The classical MoM	3.1	2.3×10^5	-1.9×10^8
	The extended MoM	7.3	−8.0	5.8×10^3
	MT	4.7	−14.2	1.1×10^4
	Our proposed	5	−11.5	8.7×10^3
R3	The classical MoM	1.2	1.2×10^5	-1.7×10^8
	The extended MoM	1.3	−30.8	4.2×10^4
	MT	2.2	−3.7	3.9×10^3
	Our proposed	1.6	−9.4	1.2×10^4
R4	The classical MoM	0.2	7.0×10^4	-1.2×10^8
	The extended MoM	0.4	−3.9	5.1×10^3
	MT	13.8	−1.6	1.2×10^3
	Our proposed	2.0	−2.4	2.4×10^3

Table 2.8 Parameter estimation results in four regions of Fig. 2.12

Regions	Methods	Parameters		
		n	α	γ
R1	The classical MoM	3.2	9.3×10^4	-1.1×10^8
	The extended MoM	4.1	−21.2	2.3×10^4
	MT	Unsolved	Unsolved	Unsolved
	Our proposed	3.8	−28.3	3.1×10^4
R2	The classical MoM	2.2	8.5×10^4	-1.7×10^8
	The extended MoM	3.5	−9.9	1.8×10^4
	MT	Unsolved	Unsolved	Unsolved
	Our proposed	2.6	−22.8	4.4×10^4
R3	The classical MoM	0.3	2.7×10^4	-4.7×10^7
	The extended MoM	0.4	−5.7	8.1×10^3
	MT	1.4	−1.4	875.9
	Our proposed	0.8	−3.1	3.7×10^3
R4	The classical MoM	0.1	4.0×10^4	-5.8×10^7
	The extended MoM	0.1	−7.9	9.9×10^3
	MT	2.5	−0.9	211.8
	Our proposed	0.6	−2.3	1.9×10^3

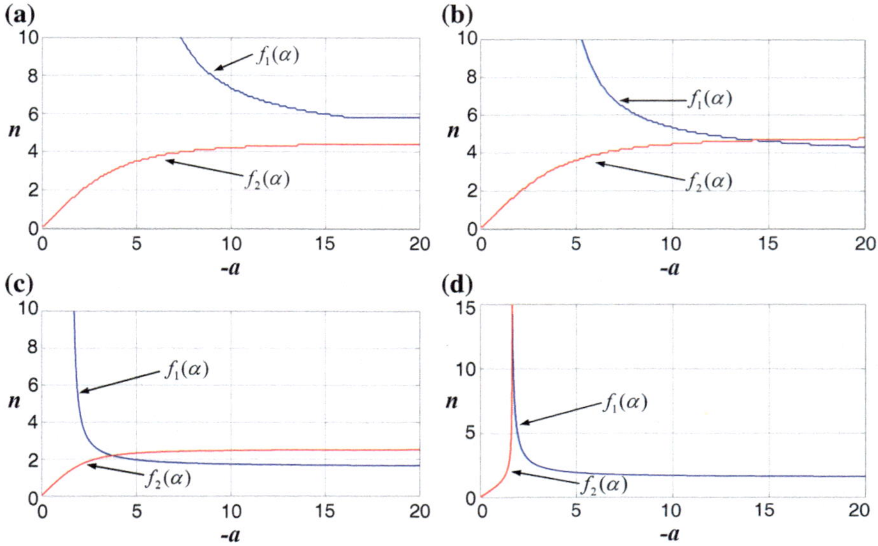

Fig. 2.14 MT based α and n estimation method by calculating the intersection point of $f_1(x)$ and $f_2(x)$ in four regions (R1 to R4) of Fig. 2.12, and the blue curve represents $f_1(\alpha)$ and the red curve represents $f_2(\alpha)$. **a** R1. **b** R2. **c** R3. **d** R4

The key-point of MT based parameter estimation is calculating the intersection point of the inverse digamma function $f_1(x)$ and $f_2(x)$. When modeling the extremely homogeneous regions, MT can't arrive at a solution, and sometime the same thing is met when modeling the homogeneous region. To deeply evaluate the MT based method, we draw the curves of $f_1(x)$ and $f_2(x)$. Figure 2.14 shows the inverse digamma function curves $f_1(x)$ and $f_2(x)$ in four regions of Figs. 2.12 and 2.15 shows the inverse digamma function curves $f_1(x)$ and $f_2(x)$ in four regions of Fig. 2.13. It can be seen that the two curves $f_1(x)$ and $f_2(x)$ don't intersect in Figs. 2.14a, and 2.15a, b. It means that no solution is arrived at, which is consistent with the MT parameters estimation results of R1 in Table 2.7, and R1 and R2 in Table 2.8.

Compared with the above three methods, the parameters obtained by our proposed method are met with the requirement of Eq. (2.60) both in Tables 2.7 and 2.8. According to literature [16] stated, the degree of heterogeneity can be measured with the estimated value of α i.e., if estimation is performed over two areas and α_1 and α_2 are the estimated parameters, then $\alpha_1 > \alpha_2$ suggests that the first area is more heterogeneous than the second. Figure 2.16 shows parameter α estimated by our method over a variety of areas (ground, vegetation, trees and buildings) in two high resolution SAR images Figs. 2.12 and 2.13. It can be seen that parameter α increases with degree of heterogeneity. The results of Fig. 2.16 are consistent with the statement of literature [16].

Tables 2.9 and 2.10 show the time consumed in estimating parameters in these eight regions of Figs. 2.12 and 2.13. The MT costs the longest time owing to the

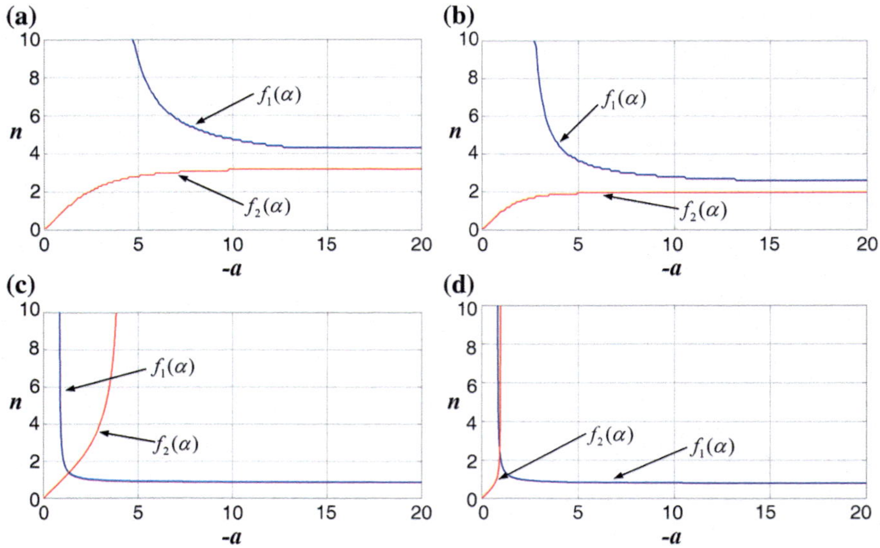

Fig. 2.15 MT based α and n estimation method by calculating the intersection point of $f_1(x)$ and $f_2(x)$ in four regions (R1 to R4) of Fig. 2.13, and the blue curve represents $f_1(\alpha)$ and the red curve represents $f_2(\alpha)$. **a** R1. **b** R2. **c** R3. **d** R4

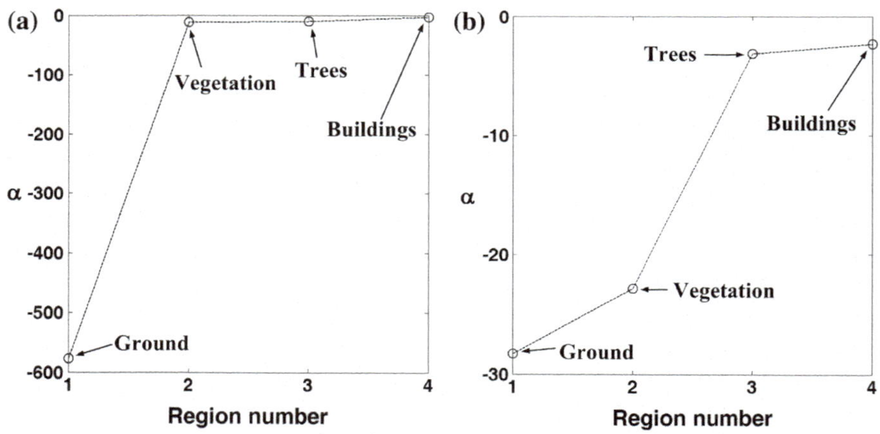

Fig. 2.16 Parameter α estimated by our method over four areas (ground, vegetation, trees and buildings) in two high resolution SAR images. **a** Four areas in Fig. 2.12. **b** Four areas in Fig. 2.13

Table 2.9 Time consuming of the four mentioned methods in four regions of Fig. 2.12

Region		R1	R2	R3	R4
Size(pixels)		235 × 235	235 × 235	235 × 235	235 × 235
Time consuming (s)	The classical MoM	0.38	0.36	0.37	0.39
	The extended MoM	0.55	0.62	0.67	0.68
	MT	2.63	3.87	3.25	3.09
	Our proposed	0.01	0.01	0.02	0.04

Table 2.10 Time consuming of the four mentioned methods in four regions of Fig. 2.13

Region		R1	R2	R3	R4
Size(pixels)		78 × 282	133 × 200	133 × 157	192 × 192
Time consuming (s)	The classical MoM	0.12	0.22	0.15	0.24
	The extended MoM	0.41	0.45	0.42	0.48
	MT	1.74	2.52	2.38	2.22
	Our method	0.01	0.01	0.01	0.02

time-consuming computation of inverse digamma function. The time-consuming of the classical MoM and the extended MoM is medial, and the extended MoM costs more time than the classical MoM because of higher order moments calculation. Our proposed is the fastest one among the four parameter estimation methods due to no higher moments calculation and inverse digamma function computation.

2.3.5.2 Fitting Precision Experiments

In order to evaluate the performance of our algorithm thoroughly, we adopt three well-known parameters like the KL distance [36], KS test [37] and MSE [35] as goodness of fit measurements.

(1) **KL Distance Measurement**

Given the theoretical PDF $p(w)$ and the estimated PDF $q(w)$, the traditional KL distance between these two densities p and q is expressed as

$$D(q\|p) = \int q(w)\log_2(\frac{q(w)}{p(w)})dw \tag{2.76}$$

Usually, the calculation of $D(q\|p)$ is given by approximate calculation:

$$D(q\|p) = \sum q(w)\Delta w \log_2(\frac{q(w)\Delta w}{p(w)\Delta w}) = \sum Q(w)\log_2(\frac{Q(w)}{P(w)}) \tag{2.77}$$

where $P(w)$ and $Q(w)$ are the values of probability. Because $D(q\|p)$ is asymmetrical, the KL distance is customarily expressed as a symmetrical form

$$D_{kl} = D(q\|p) + D(p\|q) \tag{2.78}$$

The KL distance reflects the similarity of the theoretical and estimated densities. When the estimated PDF equals the theoretical PDF, D_{kl} is zero. Otherwise, it is positive. The smaller the value of the KL distance is obtained, the higher similarity they have.

(2) **KS Test**

The KS test is based on the empirical cumulative distribution function (CDF) of the observed data. Given N observations $R_1, R_2, \ldots, R_N$ with continuous distributions, statistical goodness-of-fit is used to choose between hypothesis H_0, an assertion that the data obeys distribution q; and an alternative hypothesis H_1, an assertion that the data does not obey distribution q. The empirical CDF $\widehat{P_R}(w)$ is a piecewise constant function

$$\widehat{P_R}(w) = \begin{cases} 0, & w \to \infty \\ \frac{k}{N}, & R_k < w \le R_{k+1}, k \le N-1 \\ 1, & w > R_N \end{cases} \tag{2.79}$$

The KS test D_{ks} is defined as the supremum of the magnitude difference between the empirical CDF $\widehat{P_R}(w)$ and the cumulative distribution under hypothesis H_0, $Q_R(w)$

$$D_{ks} = \sup_w |\widehat{P_R}(w) - Q_R(w)| \tag{2.80}$$

The KS test reflects the maximal deviation between the actual CDF and hypothetical CDF. The smaller the value of the KS test is obtained, the higher goodness-of-fit they have. It is complementary for the KL distance measurement.

(3) **MSE Measurement**

Given N discrete samples $r_1, r_2, \cdots, r_N$ with the theoretical PDF $p(w)$ and the estimated PDF $q(w)$, the MSE measurement is expressed as

$$D_{mse} = \frac{1}{N}\sum_{k=1}^{N} ||p(r_k) - q(r_k)||^2 \tag{2.81}$$

where $||\cdot||$ denotes the Euclidean norm, and the estimated PDF $q(w)$ can be deduced from the histogram of the sample values. The MSE measurement reflects the mean

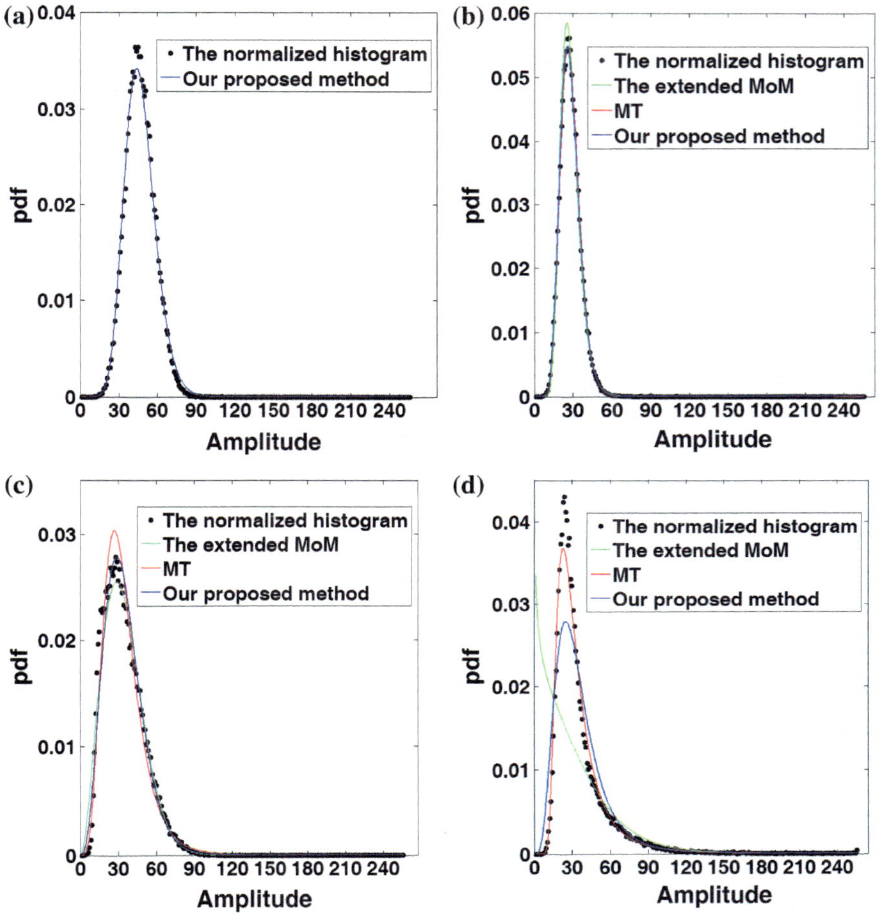

Fig. 2.17 Fitting results of four regions histogram in Fig. 2.12. **a** R1 the extremely homogeneous region. **b** R2 the homogeneous region. **c** R3 the heterogeneous region. **d** R4 the extremely heterogeneous region

deviation between theoretical PDF and the estimated PDF. The smaller the value of the MSE Measurement is, the higher goodness-of-fit they have.

Figures 2.17 and 2.18 show the fitting results of the histogram in eight regions of Figs. 2.12 and 2.13 by using the classical MoM, the extended MoM, MT and our proposed method. In Figs. 2.17 and 2.18, the horizontal axis represents the amplitude of pixel, and the vertical axis represents the PDF. Based on Eqs. (2.78), (2.80) and (2.81), the KL distance, the KS test, and the MSE measurement of the fitting result shown in Fig. 2.17 are compared in Table 2.11, and these shown in Fig. 2.18 are compared in Table 2.12. In Fig. 2.17(a), it can be seen that only the G^0 distribution density function solved by our proposed method can fit the normalized histogram of the extremely homogeneous SAR regions. The G^0 distribution density function

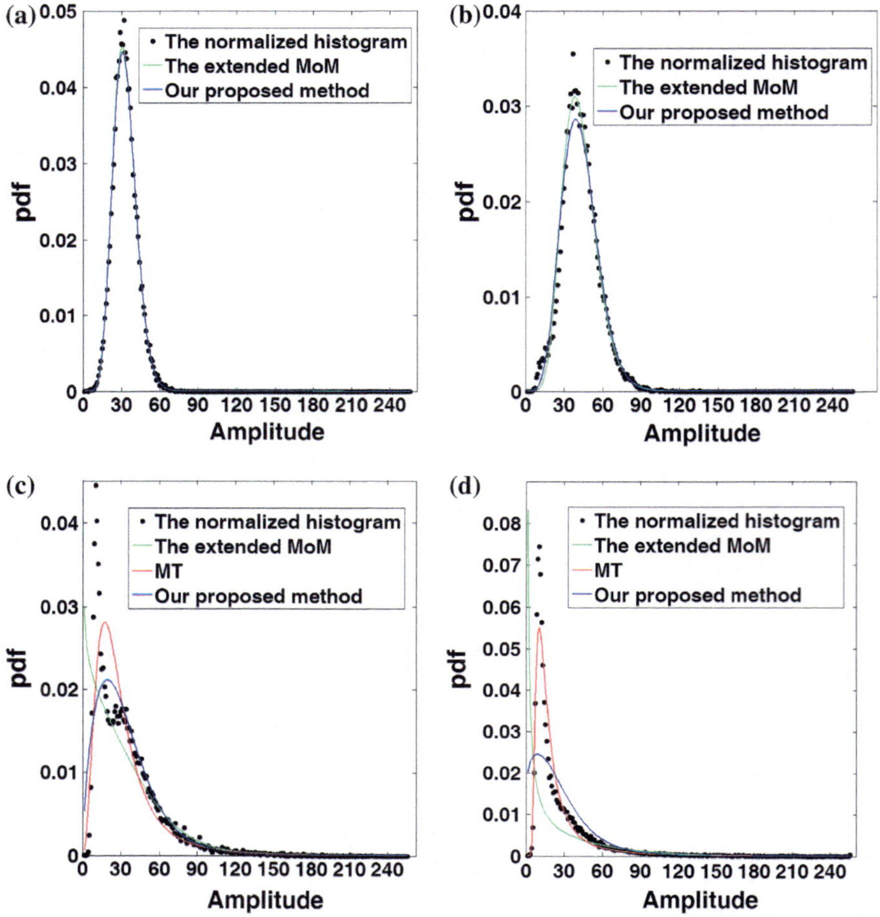

Fig. 2.18 Fitting results of four regions histogram in Fig. 2.13. **a** R1 the extremely homogeneous region. **b** R2 the homogeneous region. **c** R3 the heterogeneous region. **d** R4 the extremely heterogeneous region

solved by the extended MoM based method can model most of regions, but the fitting precisions are relatively low. The G^0 distribution density function solved by the MT based method agrees well with the histogram of the heterogeneous and extremely heterogeneous regions in Figs. 2.17d and 2.18d. Because the parameters estimated by the classical MoM method exceed the range of the requirement of Eq. (2.60), the G^0 distribution density function of these eight regions has no solution. The quantity evaluation results of goodness-of-fit in Tables 2.15 and 2.16 agree well with the fitness of the normalized histogram of these eight different degrees of heterogeneity regions in Figs. 2.17 and 2.18.

From these experimental analyses, some conclusions can be drawn

Table 2.11 The quantity comparison of fitting results of the histogram in Fig. 2.12 based on these four parameter estimation methods, where D_{kl} denotes the KL distance, and D_{ks} denotes the KS test, and D_{mse} denotes the MSE measurement

Regions	Methods	Measurements		
		D_{kl}	D_{ks}	D_{mse}
R1	The classical MoM	Unsolved	Unsolved	Unsolved
	The extended MoM	Unsolved	Unsolved	Unsolved
	MT	Unsolved	Unsolved	Unsolved
	Our proposed	0.02	0.02	1.7×10^{-9}
R2	The classical MoM	Unsolved	Unsolved	Unsolved
	The extended MoM	0.02	0.02	3.1×10^{-9}
	MT	0.00	0.00	4.5×10^{-10}
	Our proposed	0.00	0.01	7.1×10^{-10}
R3	The classical MoM	Unsolved	Unsolved	Unsolved
	The extended MoM	0.12	0.03	5.8×10^{-9}
	MT	0.05	0.03	6.4×10^{-9}
	Our proposed	0.06	0.02	5.4×10^{-9}
R4	The classical MoM	Unsolved	Unsolved	Unsolved
	The extended MoM	1.3	0.3	2.6×10^{-7}
	MT	0.03	0.02	6.0×10^{-9}
	Our proposed	0.23	0.08	3.7×10^{-8}

(1) Our parameter estimation scheme of the G^0 distribution has a wider modeling ability and lower computation time than the other three methods, but the fitting precision of which is lower than the MT based method when modeling the heterogeneous and extremely heterogeneous regions.
(2) The classical MoM based method can not model any regions of these two high resolution SAR images.
(3) The extended MoM based method is able to model most regions of these two high resolution SAR images, but the goodness-of-fit of which is the worst among these four methods.
(4) The MT based method has a higher fitting precision than the others when modeling the heterogeneous and extremely heterogeneous regions, but the calculation time of which is the longest among these four methods. Furthermore, it can not model the extremely homogeneous regions in high resolution SAR images.

Table 2.12 The fitting results of the histogram based on these four parameter estimation methods, where D_{kl} denotes the KL distance, and D_{ks} denotes the KS test, and D_{mse} denotes the MSE

Regions	Methods	Measurements		
		D_{kl}	D_{ks}	D_{mse}
R1	The classical MoM	Unsolved	Unsolved	Unsolved
	The extended MoM	0.01	0.01	3.9×10^{-9}
	MT	Unsolved	Unsolved	Unsolved
	Our proposed	0.01	0.01	4.7×10^{-9}
R2	The classical MoM	Unsolved	Unsolved	Unsolved
	The extended MoM	0.07	0.02	8.0×10^{-9}
	MT	Unsolved	Unsolved	Unsolved
	Our proposed	0.05	0.03	1.3×10^{-8}
R3	The classical MoM	Unsolved	Unsolved	Unsolved
	The extended MoM	0.87	0.15	2.7×10^{-7}
	MT	0.23	0.07	1.3×10^{-7}
	Our proposed	0.34	0.07	1.4×10^{-7}
R4	The classical MoM	Unsolved	Unsolved	Unsolved
	The extended MoM	2.72	0.33	7.9×10^{-7}
	MT	0.10	0.05	5.0×10^{-8}
	Our proposed	1.07	0.16	3.4×10^{-7}

2.4 Conclusions

In this chapter, we have developed a new statistical model, namely $G\Gamma\Gamma$, for SAR image modeling and analysis within the product model by assuming the RCS components of the return obey an empirical generalized Gamma distribution. We also demonstrate theoretically that the proposed model has the well-known $\mathcal{K}$ and $\mathcal{G}^0$ distributions as special cases. Additionally, the parameter estimators of the presented model are obtained by applying the MoLC. The experimental results show that this model provides better performance compared to $\mathcal{G}^0$ distribution, especially in the extremely heterogeneous high-resolution urban areas.

Unfortunately, the proposed model as shown in (2.6) or (2.7) is an integral representation, which may bring some limits in practice. Our future works will focus on validating this proposed model in an image processing application such as classification, filtering, detection, etc.

We have developed an empirical model, $\mathcal{HG}^{\circ}$, to exploit the knowledge of statistical characteristics of SAR amplitude or intensity images over the wide terrain classes with homogeneous, heterogeneous, and extremely heterogeneous backscattering properties. The parameter estimators of this model based on the MoLC are also provided. Consequently, we reports the performances of different land-over typologies with the $\mathcal{HG}^{\circ}$ distribution fits. The experimental results show that the $\mathcal{HG}^{\circ}$

distribution is a more advanced model compared with the known $\mathcal{G}^0$ distribution to characterize the multi-look processed SAR data.

As we known, a preliminary statistical analysis of SAR clutter data is important for designing signal processing algorithms, such as speckle filtering, target detection, building extraction, image segmentation and classification, etc. In future, it is worth of expecting to use the $\mathcal{HG}^{\circ}$ distribution in these fields. Herein, we firstly attempt to give an analytical derivation for constructing a constant false alarm rate (CFAR) detector to promote the upcoming studies of target detection in SAR images.

Given the propose amplitude model $\mathcal{HG}_A^{\circ}$ shown in (2.29), its cumulative distribution function (CDF) is written as

$$F_{\mathcal{HG}_A^{\circ}}(x) = -\frac{\sigma\lambda^{\nu\alpha+\frac{n-\alpha}{\nu}} n^{n-1}\Gamma\left(\frac{\alpha-n}{\nu}\right)}{\nu\Gamma(\nu\alpha)\Gamma(n)} x^{2n} {}_2F_1\left(\frac{\alpha-n}{\nu}, -\frac{n}{\nu}; 1-\frac{n}{\nu}; -\frac{n}{\lambda}x^{-2\nu}\right),$$
$$-\alpha, -\nu, \lambda, \sigma, n, x > 0 \tag{2.82}$$

where ${}_2F_1(\cdot\,, \cdot; \cdot; \cdot)$ is the Gauss hypergeometric function. For a given value of the false alarm probability, denoted by P_{fa}, the corresponding CFAR threshold T for the $\mathcal{HG}_A^{\circ}$ distribution can be obtained from

$$\begin{aligned}1 - P_{fa} &= F_{\mathcal{HG}_A^{\circ}}(T) \\ &= -\frac{\sigma\lambda^{\nu\alpha+\frac{n-\alpha}{\nu}} n^{n-1}\Gamma\left(\frac{\alpha-n}{\nu}\right)}{\nu\Gamma(\nu\alpha)\Gamma(n)} T^{2n} {}_2F_1\left(\frac{\alpha-n}{\nu}, -\frac{n}{\nu}; 1-\frac{n}{\nu}; -\frac{n}{\lambda}T^{-2\nu}\right)\end{aligned} \tag{2.83}$$

Considering $F_{\mathcal{HG}_A^{\circ}}(T)$ is strictly monotonously increasing, the threshold T can be accurately calculated with the help of the numerical solution or a simple bisection method [19].

Our future work will focus on demonstrating the performances of the proposed CFAR detector using some measured SAR data and investigating how the $\mathcal{HG}^{\circ}$ distribution performs when extending it to other application fields.

Aiming at having a wider modeling ability in high resolution SAR images, this chapter has proposed an improved parameter estimation scheme of the G^0 distribution, which combines the merits from the classical moment estimation and the mellin transform. Results show that our proposed method is suitable for modeling multilook clutter with widely varying degrees of homogeneity, and has a lower computation time than the other mentioned methods.

The next step is to increase the fitting precision when modeling the heterogeneous and extremely heterogeneous regions. It would also be interesting to use the proposed method for objects classification in high resolution SAR images.

References

1. C.J. Oliver, S. Quegan, *Understanding Synthetic Aperture Radar Images* (Artech House, Norwood, MA, 1998)
2. Y. Delignon, W. Pieczynski, Modelling non-Rayleigh speckle distribution in SAR images. IEEE Trans. Geosci. Remote Sens. **40**(6), 1430–1435 (2002)
3. A. Achim, E.E. Kuruoglu, J. Zerubia, SAR image filtering based on the heavy-tailed Rayleigh model. IEEE Trans. Image Process. **15**(9), 2686–2693 (2006)
4. E.E. Kuruoglu, J. Zerubia, Modeling SAR images with a generalization of the Rayleigh distribution. IEEE Trans. Image Process. **13**(4), 527–533 (2004)
5. M.S. Greco, G. Gini, Statistical analysis of high-resolution SAR ground clutter data. IEEE Trans. Geosci. Remote Sens. **45**(3), 566–575 (2007)
6. G. Moser, J. Zerubia, S.B. Serpico, SAR amplitude probability density function estimation based on a generalized Gaussian model. IEEE Trans. Image Process. **15**(6), 1429–1442 (2006)
7. A.C. Frery, H.J. Muller, C.C. Freitas Yanasse, S.J. Siqueira Sant'Anna, A model for extremely heterogeneous clutter. IEEE Trans. Geosci. Remote Sens. **35**(3), 648–659 (1997, May)
8. A.C. Frery, J. Jacobo-Berlles, J. Gambini, M. Mejail, Polarimetric SAR image segmentation with B-Splines and a new statistical model. Multidimension. Syst. Signal Process. **21**, 319–342 (2010)
9. A.C. Frery, A.H. Correia, C.C. Freitas, Classifying multifrequency fully polarimetric imagery with multiple sources of statistical evidence and contextual information. IEEE Trans. Geosci. Remote Sens. **45**(10), 3098–3109 (2007)
10. C.C. Freitas, A.C. Frery, A.H. Correia, The polarimetric G distribution for SAR data analysis. Environmetrics **16**(1) 13–11 (2005)
11. H. Allende, A.C. Frery, J. Galbiati, L. Pizarro, M-estimators with asymmetric influence functions: The GA0 distribution case. J. Stat. Comput. Simul. **76**(11), 941–956 (2006)
12. K.L.P. Vasconcellos, A.C. Frery, L.B. Silva, Improving estimation in speckled imagery. Comput. Stat. **20**(3), 503–519 (2005)
13. E. Parzen, On estimation of a probability density function and mode. Ann. Math. Stat. **33**(3), 1065–1076 (1962)
14. G. Gao, A parzen-window-kernel-based CFAR algorithm for ship detection in SAR Images. IEEE Geosci. Remote Sens. Lett. **8**(3), 557–561 (2011)
15. Q. Jiang, E. Aitnouri, S. Wang, D. Ziou, Automatic detection for ship target in SAR imagery using PNN-model. Can. J. Remote. Sens. **26**(4), 297–305 (2000)
16. J.M. Nicolas, F. Tupin, Gamma mixture modeled with second kind statistics: Application to SAR image processing. Paper presented at the IGARSS conference (Toronto, ON, Canada, 2002,) pp. 2489–2491
17. G. Moser, J. Zerubia, S.B. Serpico, Dictionary-based stochastic expectation-maximization for SAR amplitude probability density function estimation. IEEE Trans. Geosci. Remote Sens. **44**(1), 118–200 (2006)
18. G.J. McLachlan, D. Peel, *Finite Mixture Models* (Wiely, New York, 2000)
19. V.A. Krylov, G. Moser, S.B. Serpico, J. Zerubia, Enhanced dictionary-based SAR amplitude distribution estimation and its validation with very high-resolution data. IEEE Geosci. Remote Sens. Lett. **8**(1), 148–152 (2011)
20. V.A. Krylov, G. Moser, S.B. Serpico, J. Zerubia, Supervised highresolution dual-polarization SAR image classification by finite mixtures and copulas. IEEE J. Sel. Top. Signal Process. **5**(3), 554–566 (2011)
21. G. Gao, Statistical modeling of SAR images: A survey. Sensors **10**, 775–795 (2010)
22. V. Anastassopoulos, G.A. Lampropoulos, A. Drosopoulos, M. Rey, High resolution radar clutter statistics. IEEE Trans. Aerosp. Electron. Syst. **35**(1), 43–60 (1999)
23. H.C. Li, W. Hong, Y.R. Wu, P.Z. Fan, On the empirical-statistical modeling of SAR images with generalized gamma distribution. IEEE J. Sel. Top. Signal Process. **5**(3), 386–397 (2011)
24. E.W. Stacy, A generalization of the gamma distribution. Ann. Math. Stat. **33**(3), 1187–1192 (1962)

25. J.M. Nicolas, Introduction to second kind statistic: Application of log-moments and log-cumulants to SAR image law analysis. Trait. Signal **19**(3), 139–167 (2002)
26. S.N. Anfinsen, T. Eltoft, Application of the matrix-variate Mellin transform to analysis of polarimetric radar images. IEEE Trans. Geosci. Remote Sens. **49**(6), 2281–2295 (2011)
27. C. Tison, J.M. Nicolas, F. Tupin, H. Maitre, A new statistical model for Markovian classification of urban areas in high-resolution SAR images. IEEE Trans. Geosci. Remote Sens. **42**(10), 2046–2057 (2004)
28. T. Esch, M. Thiel, A. Schenk, A. Roth, A. Müller, S. Dech, Delineation of urban footprints from TerraSAR-X data by analyzing speckle characteristics and intensity information. IEEE Trans. Geosci. Remote Sens. **48**(2), 905–916 (2010)
29. S.I. Gradshteyn, I.M. Ryzhik, *Table of Integrals, Series, and Products*, 7th edn. (Academic Press, San Diego, CA, 2007)
30. R. Ravid, N. Levanon, Maximum-likehood CFAR for Weibull background. in *IEE Proceedeeings-F*, vol. 139, no. 3 (1992, June), pp. 256–264
31. X. Qin, S. Zhou, H. Zou, G. Gao, Statistical modeling of sea clutter in high-resolution SAR images using generalized gamma distribution. in *Proceeding of IEEE Computer Vision in Remote Sensor (CVRS)* (Xiamen, China, 2012, December), pp. 306–310
32. J. Martín-de-Nicolás, M.P. Jarabo-Amores, D. Mata-Moya, N. del-Rey-Maestre, J.L. Bárcena-Humanes, Statistical analysis of SAR sea clutter for classification purposes. *Remote Sens.* **6**(10) 9379–9411 (2014)
33. G. Gao, K. Ouyang, Y. Luo, S. Liang, S. Zhou, Scheme of parameter estimation for generalized gamma distribution and its application to ship detection in SAR images. IEEE Trans. Geosci. Remote Sens. **55**(3), 1812–1832 (2017)
34. V.A. Krylov, G. Moser, S.B. Serpico, J. Zerubia, On the method of logarithmic cumulants for parametric probability density function estimation. IEEE Trans. Image Process. **22**(10), 3791–3806 (2013)
35. G. Gao, S. Gao, K. Ouyang, J. He, G. Li, Scheme for characterizing clutter statistics in SAR amplitude images by combining two parametric models. IEEE Trans. Geosci. Remote Sens. **56**(10), 5636–5646 (2018)
36. D.J. Crisp, The state-of-the-art in ship detection in synthetic aperture radar imagery. DSTO, Dept. Defence, Australian Government, Canberra, Australia, Public Release Document DSTO-RR-0272 (2004)
37. M. Stasolla, J.J. Mallorqui, G. Margarit, C. Santamaria, N. Walker, A comparative study of operational vessel detectors for maritime surveillance using satellite-borne synthetic aperture radar *IEEE J. Sel. Topics Appl. Earth Observ. Remote Sens.* **9**(6) 2687–2701 (2016, June)
38. M. Abramowitz, I.A. Stegun, *Handbook of Mathematical Functions*, 10th edn. (Dover, New York, 1972)
39. S. Kullback, R.A. Leibler, On information and sufficiency. Ann. Math. Stat. **22**(1), 79–86 (1951)

Chapter 3
Target Detection and Terrain Classification of Single-Channel SAR Images

3.1 A CFAR Detection Algorithm for Generalized Gamma Distributed Background in High-Resolution SAR Images

With the improving imaging technology of the SAR, more and more high-resolution SAR images are obtained. Interpreting SAR images manually is a vast task and may lead to many mistakes. Therefore, it is greatly necessary to develop the corresponding automatic algorithms. Focusing on the target detection, many algorithms have been developed.

The constant false alarm rate (CFAR) algorithm is a classic one that has been adopted as the first step in many SAR automatic target recognition (ATR) systems [1, 2]. To keep the actual probability of false alarm (P_{fa}) as a given value, CFAR algorithm detects the pixel under test (PUT) by comparing it with an adaptive threshold that is generated according to the local background surrounding the PUT [3].

The precise description for the statistical characteristic of background plays a key role in the CFAR algorithm, which determines how accurately the CFAR property is maintained [3]. Only when the distribution applied describes the background well regardless its variety can the algorithm maintain a desired P_{fa} for the whole image. Two main factors exist that influence the description precision: the samples used as background and the underlying distribution for background.

In the past decades, numerous researches have been carried out on the approaches of selecting pixels of background for eliminating the influence from the non-clutter pixels or the "outliers". To dispose the effect of the target pixels near the PUT, the cell averaging CFAR was presented [1]. Considering the influence from the situations like clutter edge and multi-targets, many algorithms like the greatest of CFAR, smallest of CFAR, variability index CFAR, censoring CFAR and some other recently proposed algorithms including the robust estimation can provide good solutions [4–9].

Regarding the statistical modeling for clutter in the CFAR algorithms, the Gaussian distribution is mainly applied for SAR images of coarse resolution [1, 2]. However, with the increasing resolution of SAR images, it becomes insufficient and

G. Gao, *Characterization of SAR Clutter and Its Applications to Land and Ocean Observations*, https://doi.org/10.1007/978-981-13-1020-1_3

some non-Gaussian ones including the Weibull, K, G^0, alpha-stable and heavy-tailed Rayleigh distributions were introduced [4, 6, 7, 10–12]. Comparing to the Gaussian CFAR, these non-Gaussian CFAR algorithms generally have better performances of maintaining a constant actual false alarm rate (FAR) yet usually at the cost of some limitations. First, it becomes more difficult and time consuming to estimate the parameters of the distributions. Second, the corresponding algorithm may lack the closed-form expression for detection threshold and some numerical methods are needed that often increase the computational burden seriously.

In literature [13], a generalized gamma distribution (GΓD) is used for modeling many scenes of high-resolution SAR images and shows a better performance than many other classical parametric distributions in most cases. In this chapter, we try to overcome the two limitations above by employing the GΓD for modeling clutter in the CFAR algorithm. The GΓD and its corresponding parameter estimator are presented in Sect. 3.1.1. Then we derive a closed-form expression for the adaptive threshold of the algorithm in Sect. 3.1.2. Section 3.1.3 analyzes the performance of proposed algorithm using a real high-resolution SAR image acquired by the TerraSAR–X system.

3.1.1 Generalized Gamma Distribution and Its Estimation

The GΓD was first proposed by Stacy and has been widely applied in many fields [14]. A new GΓD which generalizes Stacy's model is given by [13]

$$f(x) = \frac{|v|\kappa^{\kappa}}{\sigma\Gamma(\kappa)}\left(\frac{x}{\sigma}\right)^{\kappa v-1}\exp\left\{-\kappa\left(\frac{x}{\sigma}\right)^{v}\right\}, \quad \sigma, |v|, \kappa, x > 0 \tag{3.1}$$

where σ, v and κ refer to the scale, power and shape parameters respectively. This distribution is a versatile model that has many well-known distributions like, for instance, Rayleigh, Exponential, Weibull, Gamma and inverse gamma distributions as special cases and the Log-normal distribution a limiting case [13].

The precise parameters estimation is a key factor for the practical application of a distribution. For the GΓD, the method of log-cumulants (MoLC), proposed by Nicolas for distributions defined on R^+, is a feasible approach [13, 15]. This method, with a similar idea of the method of moments, estimates the parameters by solving a system of equations of log-cumulants statistics (for detailed definitions, see [15]). The first three log-cumulants of GΓD are calculated as

$$\kappa_1 = \ln\sigma + (\Psi(\kappa) - \ln\kappa)/v, \tag{3.2}$$

$$\kappa_2 = \Psi(1, \kappa)/v^2, \tag{3.3}$$

$$\kappa_3 = \Psi(2, \kappa)/v^3, \tag{3.4}$$

where $\Psi(x) = \frac{d}{dx}\log\Gamma(x)$ and $\Psi(n, x) = \frac{d^n}{dx^n}\Psi(x)$ refer the digamma function and the nth order polygamma function respectively. Combining (3.3) and (3.4), we have a monotone function of κ as $\kappa_2^3/\kappa_3^2 = \Psi^3(1, \kappa)/\Psi^2(2, \kappa)$, which can be solved by simple numerical procedures. In [13], by using the second order approximation of the functions of $\Psi(1, \kappa)$ and $\Psi(2, \kappa)$, a analytical approximate expression for estimating κ is presented. According to (3.2)–(3.4) and the fact that $\Psi(2, \kappa) < 0$, the other two parameters are estimated by

$$v = \mathrm{sgn}(-\kappa_3)\sqrt{\Psi(1, \kappa)/\kappa_2} \tag{3.5}$$

$$\sigma = \exp\{\kappa_1 - (\Psi(\kappa) - \ln\kappa)/v\} \tag{3.6}$$

where $\mathrm{sgn}(\cdot)$ is the sign function. For the observed data set of N_C samples, noted by $\mathbf{X} = \{X_1, X_2, \ldots, X_{N_C}\}$, the first three empirical log-cumulants are computed as $\tilde{\kappa}_1 = \frac{1}{N_C}\sum_{i=1}^{N_C}\ln X_i$, $\tilde{\kappa}_2 = \frac{1}{N_C}\sum_{i=1}^{N_C}\left(\ln X_i - \hat{\kappa}_1\right)^2$ and $\tilde{\kappa}_3 = \frac{1}{N_C}\sum_{i=1}^{N_C}\left(\ln X_i - \hat{\kappa}_1\right)^3$ [13].

3.1.2 CFAR Algorithm Using GΓD for Background

The CFAR algorithm performs target detection by comparing the PUT to an adaptive threshold to maintain a constant FAR. Three important parts are there in most CFAR algorithms, the background selection, parameters estimation for the underlying background distribution and calculation of detection threshold. Figure 3.1 illustrates the detection flow of the CFAR detection algorithms.

The adaptive threshold T is determined using local background clutter around the PUT in the sliding window. Let $f(x)$ be the probability density function (PDF) estimated from Part II and $F(x)$ be the corresponding cumulative distribution function (CDF), then, for a desired FAR P_{fa}, T can be obtained from [6, 11, 12]

$$P_{fa} = \int_T^{\infty} f(x)dx = 1 - F(T). \tag{3.7}$$

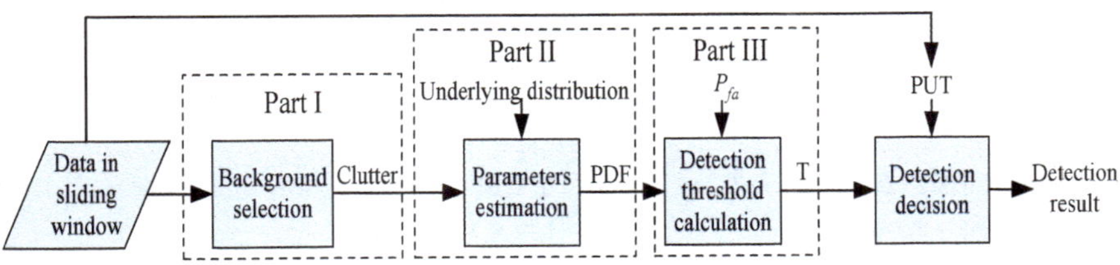

Fig. 3.1 Flow of CFAR detection algorithm

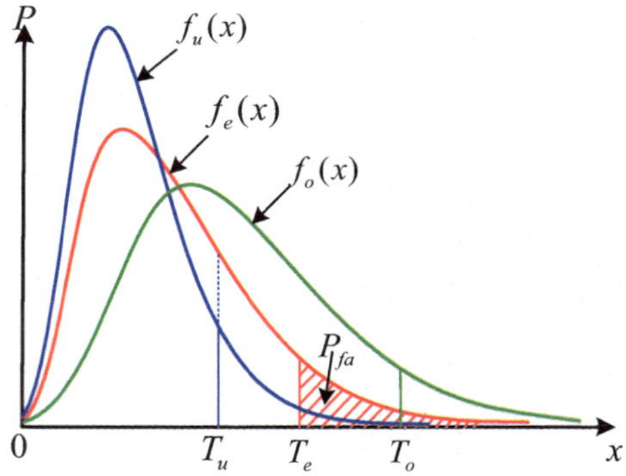

Fig. 3.2 Illustration the meaning of (3.7) and the influence of employed distribution to actual FAR

The meaning of (3.7) and the influence of the employed distributions to the actual FAR are illustrated in Fig. 3.2.

Let $f_e(x)$ be a model that fits the histogram of data exactly. Then, for a given P_{fa}, the threshold T_e generated by the CFAR algorithm with $f_e(x)$ for background leads to an actual FAR equal to the desired one. However, for the algorithms with the distributions $f_u(x)$ and $f_o(x)(f_u(T_e) < f_e(T_e) < f_o(T_e))$ for clutter, an underestimated threshold T_u and an overestimated threshold T_o are obtained that result in a larger and a smaller actual FAR than the designed one, respectively.

For the GΓD, let $y = \kappa(t/\sigma)^v$, then we have its CDF as

$$F(x) = \begin{cases} \frac{1}{\Gamma(\kappa)} \int\limits_0^{\kappa(\frac{x}{\sigma})^v} y^{\kappa-1} \exp\{-y\} dy, \; v > 0 \\ \frac{1}{\Gamma(\kappa)} \int\limits_{\kappa(\frac{x}{\sigma})^v}^{\infty} y^{\kappa-1} \exp\{-y\} dy, \; v < 0 \end{cases} \tag{3.8}$$

Using the definition of the incomplete gamma function [16]

$$P(x, a) = \frac{1}{\Gamma(a)} \int\limits_0^x t^{a-1} \exp(-t) dt, \; a, x > 0 \tag{3.9}$$

One can obtain that

$$F(x) = \begin{cases} P(\kappa(x/\sigma)^v, \kappa), & v > 0 \\ 1 - P(\kappa(x/\sigma)^v, \kappa), & v < 0 \end{cases} \tag{3.10}$$

By plugging (3.10) into (3.7), we have the adaptive threshold of CFAR algorithm for GΓD background as

$$T = \begin{cases} \sigma\kappa^{v}\{P_{Inv}(1 - P_{fa}, \kappa)\}^{1/v} & v > 0 \\ \sigma\kappa^{v}\{P_{Inv}(P_{fa}, \kappa)\}^{1/v} & v < 0 \end{cases} \tag{3.11}$$

where $P_{Inv}(x, a)$ denotes the inverse incomplete gamma function [17, 18], namely the inverse function of $P(x, a)$, which has been included as a basic function in the mathematic software Matlab.

3.1.3 Performance Evaluation

A VV-polarized X-band amplitude TerraSAR-X image of 3000×5000 pixels as shown in Fig. 3.3 is employed in this study. The image was acquired in December 2010 over the area of Hong Kong, of which the resolution is 1.9 m in range and 3.3 m in azimuth. In order to assess the contribution from the underlying distribution to the CFAR algorithm, the influence from other Parts (shown in Fig. 3.1) should be isolated. Therefore, pure data of background without target is helpful for the experiments. To this aim, a piece of pure sea clutter image of 2000×2000 pixels indicated by the rectangle in Fig. 3.3 is used here.

The procedure of our experiment to assess the performance of maintaining the FAR, which is named as CFAR maintaining performance in the letter, is summarized as follows.

(1) Clutter obtaining. The pure sea clutter image above is divided into a number of subimages with the size of $L_1 \times L_2$ pixels, each of which is assumed as the background clutter provided by the Part I of the CFAR algorithm in Fig. 3.1.
(2) Target detection. For a given P_{fa}, global detection thresholds are determined in each subimage applying the CFAR algorithms with GΓD as well as, for comparison, Gaussian, Weibull, K_A and G_A^0 distributions [19] as underlying distribution. Pixels in the subimage are compared with the global thresholds and those larger than the thresholds are treated as targets.

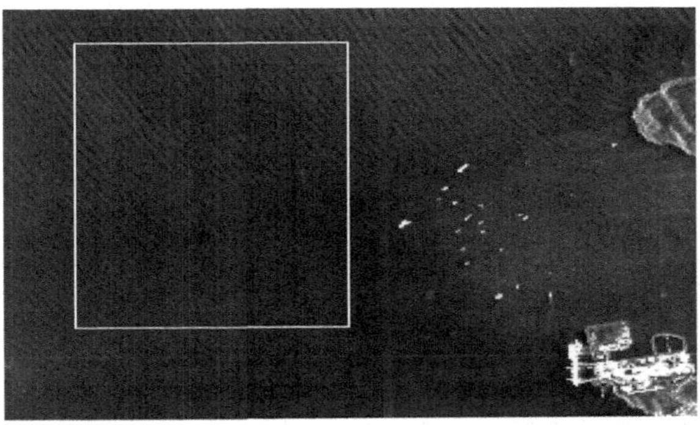

Fig. 3.3 SAR images for experiments

In the experiment, the maximum likelihood method is used for the Gaussian distribution and the MoLC for other models. (For detailed MoLC of the Weibull, K_A and G_A^0 distributions, the readers are referred to [13].) Some details of the employed distributions including the PDFs, CDFs and thresholds of the corresponding CFAR algorithms are listed in Table 3.1, where thresholds of K_A and G_A^0 distributed algorithms are calculated according to (3.5) by the numerical integral and the bisection method with the precision as 10^{-6}.

(3) Actual FAR computation. For a certain CFAR algorithm, the actual FAR of the ith subimage, noted by $R_{fa}^{(i)}$, is $R_{fa}^{(i)} = N_T^{(i)}/(L_1 \times L_2)$, where $N_T^{(i)}$ is the number of "targets" detected in the corresponding subimage.

(4) Performance comparison. The CFAR algorithm has a better CFAR maintaining performance if the corresponding actual FARs match the given P_{fa} better.

In our experiment, the pure sea clutter image above is divided into 100 subimages (200×200 pixels for each one). For three different given P_{fa} as 0.0100, 0.0200 and 0.0300, the actual FARs obtained on each subimage by all the employed CFAR algorithms are shown in Fig. 3.4, and the corresponding means are listed in Table 3.2. Besides, in order to assess the fitting results, the Kolmogorov-Smirnov (KS) distance [20] is adopted, of which a smaller value implies a better fit of the distribution to the data. For each distribution, the mean KS distance to one subimage is listed in Table 3.2 too.

It can be seen from Fig. 3.4 that the actual FARs obtained by the K_A, G_A^0 and GΓ distributed algorithms are approximate to the given values, more than that by the Gaussian and Weibull distributed algorithms, and that by the proposed algorithm match the given values best in general. The mean KS distances in Table 3.2 show that the GΓD fits the data best generally, follow the K_A, G_A^0, Weibull and Gaussian distributions. This result is consistent with the fact that the mean actual FARs obtained by the proposed algorithm match the given values best. Therefore, the proposed algorithm has a better CFAR maintaining performance than the compared ones to the data in the experiment.

Moreover, the relative magnitude of actual FARs obtained by different algorithms (shown in Fig. 3.4) can be explained by a representative fitting result as illustrated in Fig. 3.5, where the relative position of the tail-parts of PDFs can be discriminated clearly. According to the analysis in Sect. 3.1.2, one can obtain that, for a small given P_{fa}, the Gaussian, Weibull, GΓ, K_A and G_A^0 distributed algorithms would generate thresholds with magnitudes from small to large that result in actual FARs of a inverse order from large to small.

In fact, the algorithm would have a better CFAR maintaining performance in case that the corresponding distribution fits the background better. For many scenes of high-resolution SAR images, it has been shown that the GΓD provides a better goodness-of-fit than many other parametric distributions [13]. Consequently, a better CFAR maintaining performance the proposed CFAR algorithm would have in most cases.

For many scenes of high-resolution SAR images, K_A, G_A^0 and GΓD can provide well goodness-of-fit. In this section, we compare the efficiency of the CFAR algo-

Table 3.1 The employed distributions (PDFs and CDFs) and the corresponding CFAR Algorithms' thresholds

Model	PDF	CDF	Threshold "T"
Gaussian	$f(x) = \frac{1}{2\pi\sqrt{\sigma}} \exp\left\{-\frac{(x-\mu)^2}{2\sigma^2}\right\},\ \sigma, \mu, x > 0$	$F(x) = \Phi\left(\frac{x-\mu}{\sigma}\right)$	$T = \sigma\Phi^{-1}(1 - P_{fa})x + \mu$
Weibull	$f(x) = \frac{C}{B}\left(\frac{x}{B}\right)^{C-1} \exp\left(-\left(\frac{x}{B}\right)^C\right),\ B, C, x > 0$	$F(x) = 1 - \exp\left\{-\left(\frac{x}{B}\right)^C\right\}$	$T = B(-\log P_{fa})^{1/C}$
K_A	$f(x) = \frac{4\lambda n x}{\Gamma(\alpha)\Gamma(n)}\left(\lambda n x^2\right)^{\frac{\alpha+n}{2}-1} K_{\alpha-n}\left(2x\sqrt{\lambda n}\right),\ \alpha, \lambda, n, x > 0$	$F(x) = \frac{4(\lambda n)^{(\alpha+n)/2}}{\Gamma(\alpha)\Gamma(n)} \cdot \int_0^x t^{\alpha+n-1} K_{\alpha-n}\left(2t\sqrt{\lambda n}\right)dt$	Solve T from (7) by numerical method.
G_A^0	$f(x) = \frac{2n^n \Gamma(n-\alpha)\gamma^{-\alpha}}{\Gamma(n)\Gamma(-\alpha)} \frac{x^{2n-1}}{(\gamma+nx^2)^{n-\alpha}},\ -\alpha, \gamma, n, x > 0$	$F(x) = \frac{n^{n-1}\Gamma(n-\alpha)x^{2n}}{\gamma^n \Gamma(n)\Gamma(-\alpha)} {}_2F_1\left(n-\alpha, n; n+1; -\frac{n}{\gamma}x^2\right)$	Solve T from (7) by numerical method.
GΓD	$f(x) = \frac{\lvert\nu\rvert\kappa^\kappa}{\sigma\Gamma(\kappa)}\left(\frac{x}{\sigma}\right)^{\kappa\nu-1} \exp\left\{-\kappa\left(\frac{x}{\sigma}\right)^\nu\right\},\ \sigma, \lvert\nu\rvert, \kappa, x > 0$	$F(x) = \begin{cases} P\left(\kappa(x/\sigma)^\nu, \kappa\right), & \nu > 0 \\ 1 - P\left(\kappa(x/\sigma)^\nu, \kappa\right), & \nu < 0 \end{cases}$	$T = \begin{cases} \sigma\kappa^\nu\left\{P_{Inv}\left(1 - P_{fa}, \kappa\right)\right\}^{1/\nu}, & \nu > 0 \\ \sigma\kappa^\nu\left\{P_{Inv}\left(P_{fa}, \kappa\right)\right\}^{1/\nu}, & \nu < 0 \end{cases}$

Here $\Phi(\cdot)$ and $\Phi^{-1}(\cdot)$ are the standard normal cumulative distribution function and its inverse function, ${}_2F_1(a, b; c; \cdot)$ the hypergeometric function and $K_\alpha(\cdot)$ the modified Bessel function of the second kind of order α

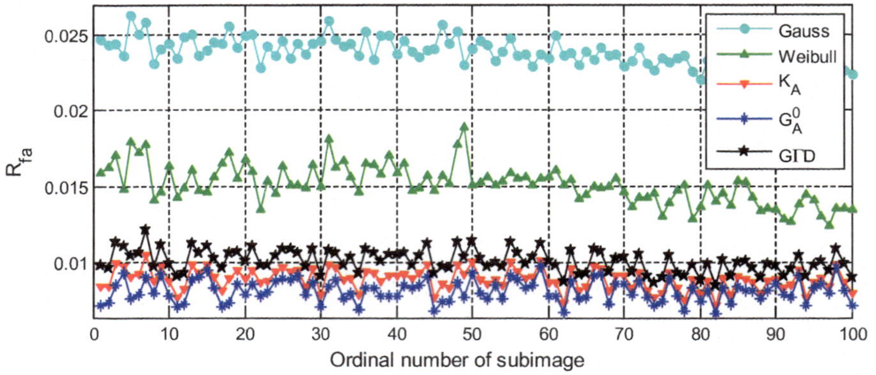

(a) $P_{fa} = 0.0100$

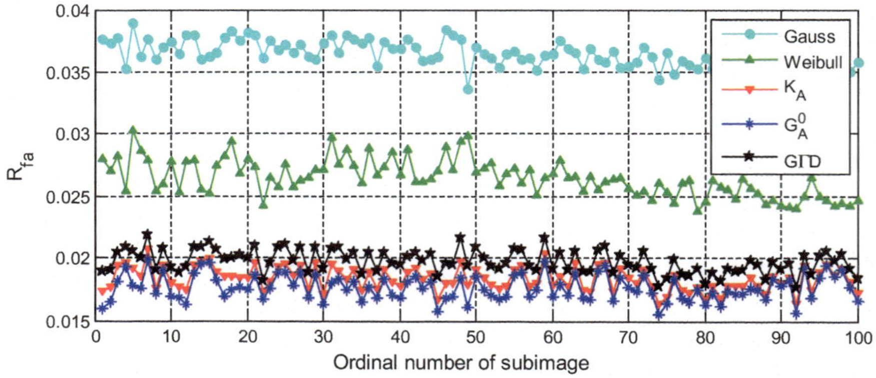

(b) $P_{fa} = 0.0200$

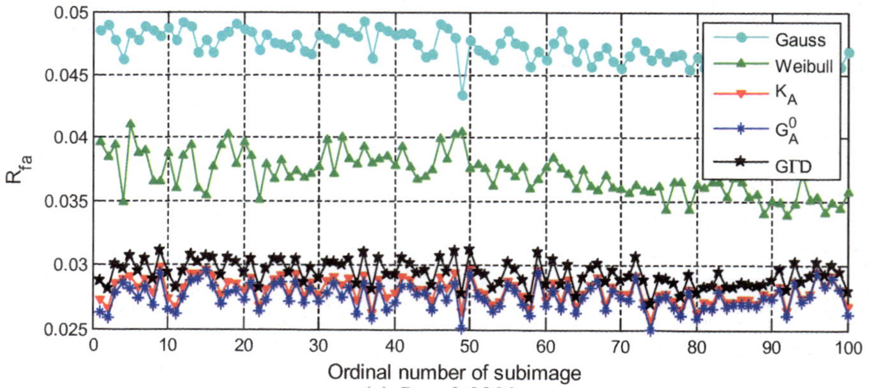

(c) $P_{fa} = 0.0300$

Fig. 3.4 Actual FARs obtained by CFAR algorithms using different distributions under three given Pfa

Table 3.2 Mean KS distances and mean actual fars corresponding in Fig. 3.4

Underlying distribution	Mean KS distance	Given P_{fa}		
		0.0100	0.0200	0.0300
		Mean R_{fa}		
Gaussian	0.0883	0.0238	0.0364	0.0472
Weibull	0.0089	0.0151	0.0264	0.0371
K_A	0.0051	0.0089	0.0183	0.0280
G_A^0	0.0057	0.0081	0.0176	0.0275
GΓD	0.0044	0.0101	0.0197	0.0294

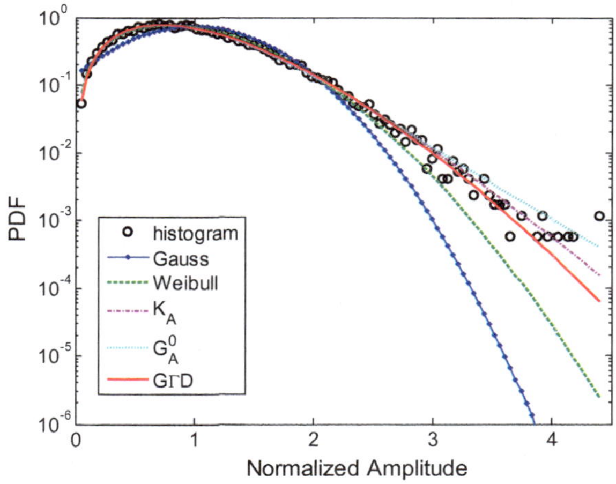

Fig. 3.5 A representative fitting result

rithms with these distributions for background. The experiments are carried out by Matlab codes running on the same hardware platform of Pentium (R) Dual-Core 2.5 GHz CPU and 2.0 GB memory.

A detection experiment similar to that before is performed, in which the subimage size is 50×50 pixels instead of 200×200 pixels. Table 3.3 lists the time consumed on these two experiments including that for the statistics calculation, estimation equations solution and the threshold calculation. The total time needed at least for each CFAR algorithm is the sum of the least time of each part.

From Table 3.3 the following conclusions can be drawn, i.e.,

(1) For estimating parameters of the three distributions, the times cost by the original MoLC that solves the estimation equations by numerical methods are of the same order of magnitude, while nearly two orders of magnitude larger than that cost by the analytical approximate MoLC for GΓD.
(2) Referring to the threshold calculation, the time cost for the proposed algorithm is about two and four orders of magnitude lower than the K_A and G_A^0 distributed algorithms, respectively. For these two compared algorithms, numerical inte-

Table 3.3 Mean time cost of three different CFAR algorithms on data of different sizes

Underlying distribution	Size of subimage (pixels)	Statistics calculation (ms)	Estimation equations solution (ms)		Threshold calculation (ms)	Total time at least (ms)
			Original MoLC	Analytical MoLC		
K_A	200 × 200	8.8902	10.9302	/	2209.1678	2228.9882
G_A^0			11.4855	/	6.3511	26.7267
GΓD			7.0089	0.0802	0.0955	9.0659
K_A	50 × 50	0.9420	12.2001	/	1885.0597	1898.2018
G_A^0			15.0734	/	8.8915	24.9069
GΓD			6.5207	0.0905	0.1021	1.1346

gral and numerical method (bisection method here) are applied, which are usually time consuming. Moreover, during the procedure of bisection method, the numerical integral to compute CDF performs many times that have significantly increased the number of operations. Besides, due to complicated density and CDF of the K_A distribution relying on the modified Bessel function of the second kind, the K_A distributed algorithm costs much more time than the G_A^0 distributed one.

(3) Regardless of the algorithm, time cost on the estimation equation solution and threshold calculation almost keep on the same level, while that on the statistics calculation (which are the same for all the algorithms) decreases with the decreasing subimage sizes. Moreover, as can be seen that the ratio of time on statistics calculation to the total time is, take the subimage of 200 × 200 pixels for an example, 98.06% for the proposed algorithm, while that for the K_A and G_A^0 distributed algorithms are 0.40 and 33.26% respectively. Accordingly, the proposed algorithm is much more efficient than the two compared ones especially when the size of the data is small (like 50 × 50 pixels) or the time for statistics calculation is cut down by fast algorithms as that in [6].

3.2 A Parzen Window Kernel Based CFAR Algorithm for Ship Detection in SAR Images

Ship detection in SAR images is an important application of earth observation for monitoring of fishing vessels, oil pollution and warship reconnaissance [21, 22]. Many investigations have been carried out in the literature, such as CFAR (Constant False Alarm Rate) [23], the wavelet transform-based [21], and coherence images-based [22], etc. Among these algorithms, CFAR detection, which is famous for its constant false alarm probability and adaptive threshold, has been used widely. The precision of statistical modeling for clutter is crucial for the detection performance.

Presently, the conventional CFAR detections mainly adopt parametric models, such as Gamma, Weibull and K etc. [23, 24]. However, the modeling ability of a parametric model is doubtable for a complex unknown SAR scene.

In fact, the nonparametric models are more flexible and can more accurately fit the real data [25–27]. The Parzen window kernel method [25–27] is a commonly used nonparametric probability density estimation strategy in the field of pattern recognition. It is a kind of non-parametric method and a data-driven model to estimate the probability density function (PDF) of SAR image data with excellent estimation accuracy. The Parzen window kernel method is suitable for estimating the complex unknown PDF. However, the acquirement of detection threshold is a hard task for CFAR algorithm based on the Parzen window kernel. In this chapter, a numerical solution of the threshold will be presented and tested over real data.

3.2.1 *Statistical Modeling of SAR Image Based on Parzen Window Kernel*

The basic idea of the Parzen window kernel method is to utilize the weighted sum of different kernel functions for obtaining the estimation of the statistical distribution. Commonly used kernel functions include the uniform, triangle, cosine and Gaussian. In this work, we used the standard normal distribution as the kernel function:

$$\varphi(u) = \frac{1}{\sqrt{2\pi}} \exp\left(-\frac{u^2}{2}\right) \tag{3.12}$$

The corresponding cumulative distribution function is given by

$$\Phi(u) = \frac{1}{\sqrt{2\pi}} \int_{-\infty}^{u} \exp\left(-\frac{t^2}{2}\right) dt \tag{3.13}$$

Therefore, the estimation of PDF for SAR image follows the approximation of the kernel functions as below

$$\hat{p}_N(x) = \frac{1}{N} \sum_{j=1}^{N} \frac{1}{h_N} \varphi\left(\frac{x - x_j}{h_N}\right) \quad x \geq 0 \tag{3.14}$$

where $x_1, x_2, \ldots, x_N$ denote the samples and are corresponded to the value of pixels in SAR image. N represents the number of sample points. $h_N\,(h_N > 0)$ is the bandwidth that indicates the width of the kernel function. From Eq. (3.14), the Parzen window kernel method is actually a mixed distribution by accumulating different kernel functions. The estimation expression of image PDF is obtained by the weighted sum of the kernel functions in samples. Therefore, the characteristic of this method is suitable for the estimation of various complex and unknown PDF, in spite of single-peak, multi-peaks, regulation or non-regulation.

In Eq. (3.14), small h_N will make the PDF estimate appear noisy and show spurious features while big one will lead to smooth estimates where important structural features may be missed [25, 26]. The selection of bandwidth h_N can adopt several methods, such as the plug-in estimators and data driven manners [25, 26]. On the other hand, h_N should decrease with the increase of N so as to make $\hat{p}_N(x)$ be convergence. In this study, we used the method in literature [27] to set a fixed value of h_N, which is defined as,

$$h_N = h\Big/\sqrt{N} \tag{3.15}$$

where h is an adjustable constant. Commonly, h is an integer and determined empirically.

3.2.2 CFAR Detection

Assuming the PDF estimation of SAR image is $\hat{p}_N(x)$, we combine Eq. (3.14) under the condition that the theoretical false alarm probability of detection is p_{fa}, then, the global detection threshold T is given by

$$\begin{aligned} p_{fa} &= \int_T^{\infty} \hat{p}_N(x)dx \\ &= \frac{1}{h_N}\frac{1}{N}\sum_{j=1}^{N}\int_T^{\infty}\frac{1}{\sqrt{2\pi}}\exp\left[-\frac{1}{2}\left(\frac{x-x_j}{h_N}\right)^2\right]dx \end{aligned} \tag{3.16}$$

Let $t = \frac{x-x_j}{\sqrt{2}h_N}$, then,

$$p_{fa} = \frac{1}{N}\frac{1}{\sqrt{\pi}}\sum_{j=1}^{N}\int_{\frac{T-x_j}{\sqrt{2}h_N}}^{\infty} e^{-t^2}dt \tag{3.17}$$

According to the error function, and the complementary error function whose expression is $erfc(x) = 1 - \frac{2}{\sqrt{\pi}}\int_0^x e^{-t^2}dt$, the Eq. (3.17) can be written by

$$p_{fa} = \frac{1}{2N}\sum_{j=1}^{N} erfc\left(\frac{T-x_j}{\sqrt{2}h_N}\right) \tag{3.18}$$

The detection threshold T can be finally determined by Eq. (3.18). Thus, for a pixel, if its value exceeds T, this pixel is considered to be a target point; otherwise, it is declared to be a clutter point. The target detection is completed by comparing all pixels in the tested SAR image with T.

Considering Eq. (3.18), the relation of detection threshold T and false alarm probability is included in the accumulation of N complementary error functions. T is very difficult to derive theoretically given p_{fa}. Consequently, the numerical solution of T is requested to be devised.

According to Eq. (3.14), the CDF (cumulative distribution function) of $\hat{p}_N(x)$ can be obtained by

$$F_N(x) = \int_0^x \hat{p}_N(t)dt \tag{3.19}$$

Moreover, the relationship of $F_N(T)$ and p_{fa} can be expressed as

$$p_{fa} = \int_T^\infty \hat{p}_N(x)dx = 1 - F_N(T) \tag{3.20}$$

Let $\mathbf{X}$ represents the actual SAR image, then,

$$\mathbf{X} = \{x_1, x_2, \ldots, x_N\} \tag{3.21}$$

If $x_{\max}$ denotes the maximum value of the samples, i.e.

$$x_{\max} = \max(\mathbf{X}) \tag{3.22}$$

We divide the range $[0, x_{\max}]$ into equal intervals according to the step-length τ and define the following equation:

$$\mathbf{X}' = \left\{x_{(1)}, x_{(2)}, \ldots, x_{(m)}\right\} \tag{3.23}$$

where $x_{(l)}, 1 \le l \le m$ indicate the nodes of different intervals, and $x_{(1)} = x_1$. The number of nodes is given by

$$m = x_{\max}/\tau + 1 \tag{3.24}$$

Because $T \in [0, x_{\max}]$, the selection of τ is determined by the expected precision of T. For example, if the expected precision of T does not exceed 0.01, $\tau = 0.01$. We compute the CDFs $F_N\left(x_{(l)}\right), 1 \le l \le m$ separately, then $\mathbf{F'_N} = \left\{F_N\left(x_{(1)}\right), F_N\left(x_{(2)}\right), \ldots, F_N\left(x_{(m)}\right)\right\}$, where the expression of the numerical solution for $F_N\left(x_{(l)}\right)$ is given by

$$F_N\left(x_{(l)}\right) = \begin{cases} \sum_{k=1}^{l-1} \left(\frac{\hat{p}_N\left(x_{(k)}\right)+\hat{p}_N\left(x_{(k+1)}\right)}{2} \cdot \tau\right), & m \geq l \geq 2 \\ 0, & l = 1 \end{cases} \tag{3.25}$$

Since the computation of the CDF $F_N(\cdot)$ is a process of accumulation, the iterative relationship can be given by

$$F_N\left(x_{(l)}\right) = F_N\left(x_{(l-1)}\right) + \frac{\hat{p}_N\left(x_{(l-1)}\right) + \hat{p}_N\left(x_{(l)}\right)}{2} \cdot \tau, \quad m \geq l \geq 2 \tag{3.26}$$

Assuming $\mathbf{Q_N}$ is the complementary of $\mathbf{F'_N}$, i.e.

$$\mathbf{Q_N} = 1 - \mathbf{F'_N} = \{Q_N(1), Q_N(2), \ldots, Q_N(m)\} \tag{3.27}$$

Obviously, $\mathbf{Q_N}$ is monotonously degressive and $Q_N(l) = 1 - F_N\left(x_{(l)}\right)$.

Summing up, Fig. 3.6 shows the detailed flow chart of calculating the detection threshold, which can be divided as 7 steps:

Step 1. Determining the step-length τ by the expected precision of T, and giving false alarm probability p_{fa};
Step 2. Determining $x_{\max}$, $\mathbf{X}'$ and m separately by Eqs. (3.22), (3.23) and (3.24);
Step 3. Calculating $\hat{p}_N\left(x_{(1)}\right), \hat{p}_N\left(x_{(2)}\right), \cdots, \hat{p}_N\left(x_{(m)}\right)$ by Eqs. (3.14) and (3.23);
Step 4. Computing $\mathbf{F'_N}$ by Eqs. (3.25) and (3.26);
Step 5. Let $l = 1$, calculating $\mathbf{Q_N}$ by Eq. (3.27);
Step 6 Calculating $Q_N(l+1)$ according to $Q_N(l)$. If $Q_N(l) > p_{fa} > Q_N(l+1)$, $T = x_{(l)}$ and return; otherwise, continuing to next step;
Step 7. Let $l = l + 1$, returning to Step 6.

3.2.3 Experimental Results

The proposed method is tested in real SAR images. Figure 3.7 shows a representative SAR image used in this study, the horizontal and vertical axes are the directions of range and azimuth, respectively. The data were collected by the space-borne SAR near Kaohsiung harbor in Taiwan Strait, with the size of 3325 × 2877 pixels. The image is single-look. As shown in Fig. 3.8, there are 16 ships in total, which are numbered 1–16, recorded by observers during the time of the SAR image acquisition under the condition that wind speed is less than 1.05 m/s and sea state is calm. Table 3.4 lists the information about these ships.

Figure 3.9 gives the fitting result of the histogram of Fig. 3.7 based on the Parzen window kernel and K distribution [23, 28]. In order to quantitatively assess the fitting result, we adopt KL (Kullback-Leibler) [29] distance and KS (Kolmogorov-Smirnov) [30] test as similarity measurements.

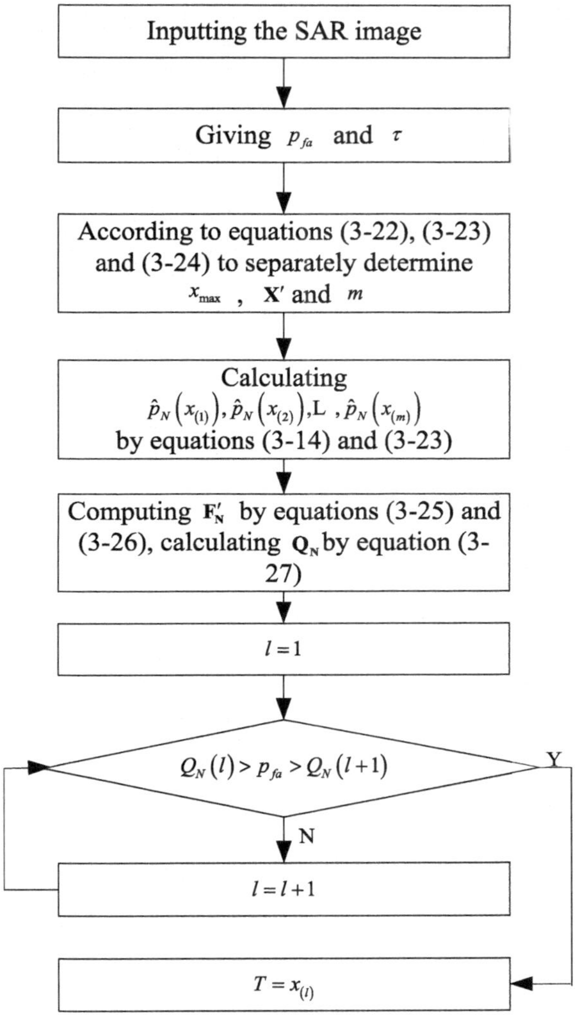

Fig. 3.6 The detailed flow of calculating the detection threshold

Obviously, KS test is complementary for KL distance measurement. Figure 3.9 shows fitting results of the histogram in Fig. 3.7 using K distribution and the Parzen window kernel. The horizontal axis marked as the amplitude of pixel and the vertical axis represents PDF in Fig. 3.9. The KS value and the KL value of the fitting result shown in Fig. 3.9 are compared in Table 3.5, where h is set empirically as 40. Thus, it is easy to find that the modeling method based on Parzen window kernel well agrees with the given SAR image, which implies the higher precision of fitting using Parzen window method than using K distribution.

With $p_{fa} = 0.01$ and $\tau = 0.1$, we compare the proposed algorithm with the CFAR using K distribution which has been successfully applied in OMW (Ocean

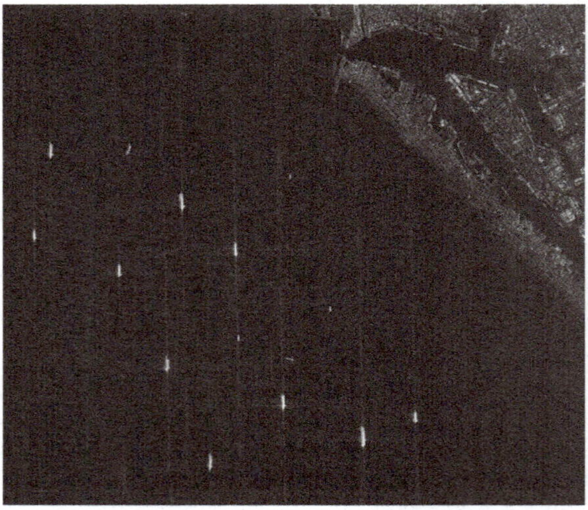

Fig. 3.7 The representative SAR image

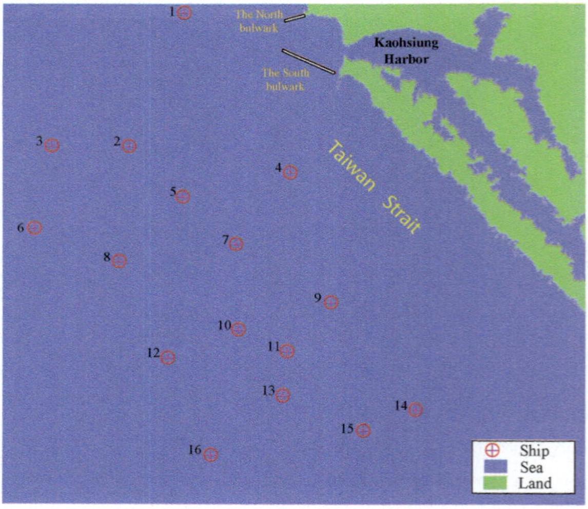

Fig. 3.8 The illustration of scene content

Monitoring Workstation) of Canada [23, 24]. Figure 3.10 shows the detection results of the two algorithms. From Fig. 3.10, there are more clutter false alarms occurred in CFAR algorithm based on K distribution than in our method under the same false alarm probability. This is because the higher fitting precision can be acquired using the Parzen window method than K distribution and the mismatch of statistical modeling results in significant CFAR loss. Utilizing non-optimized matlab7.1 codes under a hardware environment of 2G CPU and 1G memory, the time consumptions are 2.0167 and 64.3682 s, respectively subjected to CFAR algorithm using K distribution and the proposed method.

Table 3.4 The location and length of ships

No.	Longitude	Latitude	Length (m)
1	120°14′19″	22°37′39″	32.9
2	120°13′33″	22°36′29″	91.7
3	120°12′47″	22°36′35″	155.4
4	120°15′06″	22°36′00″	19.6
5	120°14′00″	22°35′56″	187.3
6	120°12′29″	22°35′50″	111.6
7	120°14′28″	22°35′25″	139.7
8	120°13′16″	22°35′23″	120.2
9	120°15′18″	22°34′44″	20.5
10	120°14′21″	22°34′35″	25.7
11	120°14′49″	22°34′19″	47.1
12	120°13′36″	22°34′27″	127.9
13	120°14′41″	22°33′55″	158.1
14	120°15′59″	22°33′36″	113.8
15	120°15′26″	22°33′31″	240.4
16	120°13′53″	22°33′29″	174.3

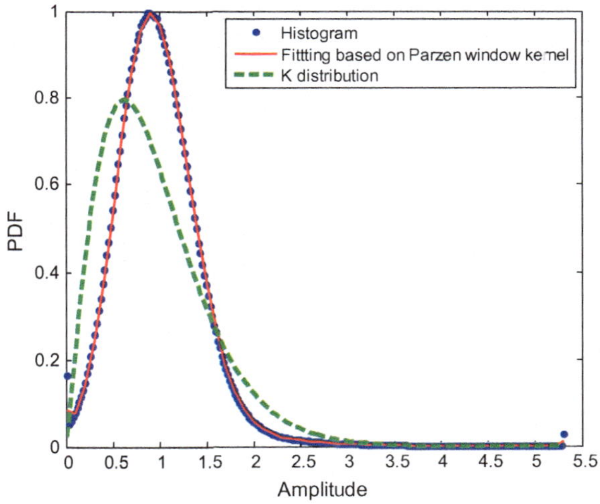

Fig. 3.9 The fitting result of histogram

Table 3.5 The comparison of fitting histogram

Measures	*K* distribution	The Parzen window kernel
KS	0.1297	0.0121
KL	0.1532	0.0016

Fig. 3.10 The detection results of ships

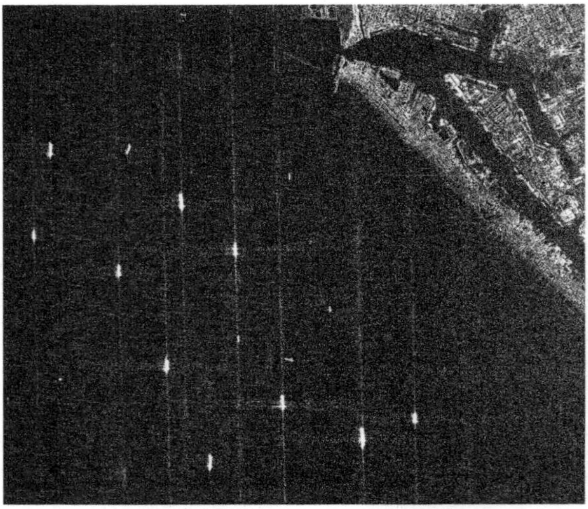

(a) the CFAR using K distribution

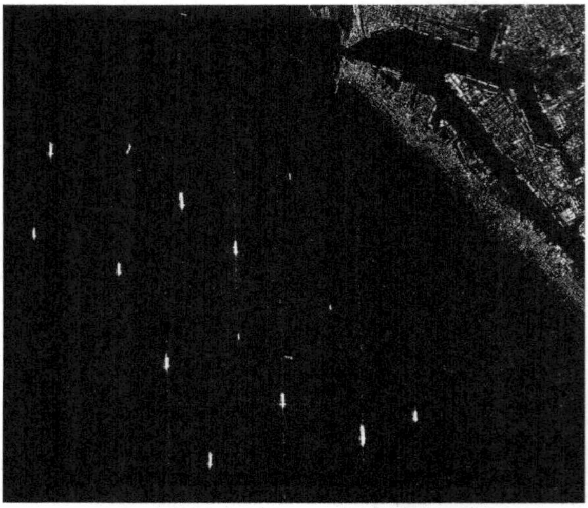

(b) the proposed method

In order to evaluate the performance of the algorithms in a quantitative manner, the estimation of the target detecting probability is defined as

$$p_d = \frac{N_{td}}{N_{total\_target}} \tag{3.28}$$

where N_{td} is the number of target pixels detected, N_{total_target} denotes the number of all target pixels. N_{total_target} is determined by a manual segmentation result of targets.

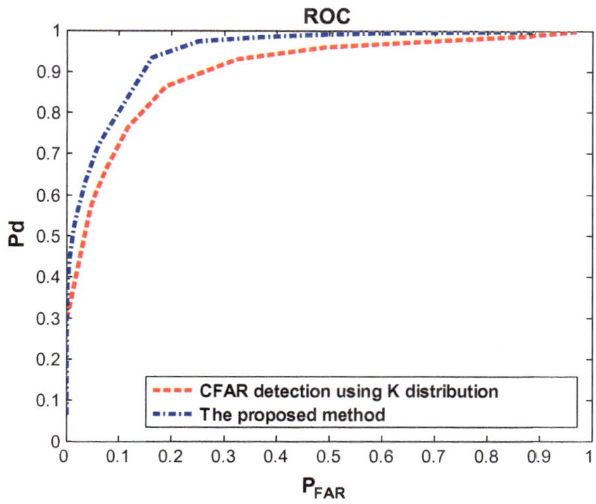

Fig. 3.11 The comparison of Performance curves for detection

Meanwhile, in order to remove the land disturbance, we utilize Fig. 3.8 to mask Fig. 3.7 If the size of image is $N_0 \times M_0$ and the number of land pixels is N_L, then the number of sea clutter pixels can be denoted as $N_{total\_clutter}$, is $N_0 \times M_0 - N_L - N_{total\_t\arg et}$. The actual false alarm ratio is defined as

$$p_{FAR} = \frac{N_{cd}}{N_{total\_clutter}} \tag{3.29}$$

where N_{cd} represents the number of false alarm pixels generated by sea clutter.

As shown in Fig. 3.11, combining Eqs. (3.28) and (3.29), we obtain the receiver operating characteristic (ROC) curves of the CFAR algorithm using K distribution and the presented method. These curves indicate that the method in this study has better detection precision than the CFAR algorithm based on K distribution.

3.3 A Markovian Classification Method for Urban Areas in High-Resolution SAR Images

In the last few years, high quality images of the earth produced by SAR systems carried on a variety of airborne and spaceborne platforms have become increasingly available [31]. Thanks to high resolution, many classes of objects are visible in images. This information is of high importance for equipment planning, natural risk prevention, and defense.

In the context of SAR data classification, a crucial problem is represented by the need to develop accurate models for the statistics of pixel intensities [32, 33]. We choose G^0 distribution [34], proposed by Frery in 1997, which is a general distribution for statistically modeling the homogeneous, heterogeneous and extremely heterogeneous clutter regions. Unfortunately, classical statistical parameter estimation techniques (Maximum likelihood (ML), moment estimations) fail for G^0 distributions because some of its moments do not exist and ML method is rather complicated. Therefore, we make use of a recent method which based on Mellin transform, proposed by Nicolas in [35].

In order to preserve spatial coherence, a Markovian classification [36] is adopted. For a specifically image, each class is associated with a statistical PDF by the training process beforehand. Then, the classification problem can be addressed.

3.3.1 Markovian Formalism

The statistical model is now validated, and we can expect to have accurate estimated PDFs to describe each class. We propose to use this statistical information in a Bayesian classification scheme and to improve the classification by using the spatial context in a Markovian framework. The Markovian formalism is first detailed; then, the algorithm steps of the classification method are explained.

Let X be a field of random variable and X_s the value of pixel s. A label l_s is associated to each pixel, and the set of labels defines the unknown realization of the label random field l.

Assuming pixel-to-pixel and pixel-to-label independence, using the Hammersley-Clifford theorem [36], the conditional probability of having a label field l given a radiometric field x can be written under Gibbs field formalism as

$$P(L = l|X = x) = \frac{1}{Z} e^{-U(l|x)/T}. \tag{3.30}$$

The Z is a normalization coefficient, and U is the energy function, which comprise two parts:

$$U(L|X) = -\sum_{S} \ln p(x_s|l_s) + \frac{1}{T} \sum_{c \in N_s} V_c(l_s, l_t) \tag{3.31}$$

where C is the set of the chosen neighborhood and Vc is the potential of the label configuration. In this paper, the model below is used.

$$Vc(l_s, l_t) = \begin{cases} -\beta, & l_s = l_t \\ \beta, & l_s \neq l_t \end{cases}. \tag{3.32}$$

The result of the Markovian classification is the field of labels L which maximizes P or, equivalently, minimizes U, it can be expressed as follows

$$X_{opt} = \arg\min(U(l|x)). \tag{3.33}$$

3.3.2 Optimization Algorithm

In order to solve the optimization problems, classical techniques such as Iterated Conditional Model (ICM) Algorithm [37] and Simulated Annealing (SA) algorithm [36] are commonly used. Here the deterministic Modified Metropolis Dynamics (MMD) algorithm [38] is suggested. MMD algorithm proceeds as follows:

(1) Sample a random initial configuration W^0, with $k = 0, T = T_0$;
(2) Using uniform distribution pick up a global state η which differ exactly in one element from W^k;
(3) Compute $\Delta U = U(\eta) - U(w^k)$ and accept according to the rule:

$$w_{i,j}^{k+1} = \begin{cases} \eta, & if\, \Delta U \le 0, \\ \eta, & if\, \Delta U \le 0 \; and \; \ln(a) \le -\frac{\Delta U}{T}, \\ w^k, & otherwise. \end{cases} \tag{3.34}$$

where a is a constant threshold ($a \in (0, 1)$), chosen at the start of the algorithm;
(4) Decrease the temperature and go to Step 2 until convergence ($T = T_{end}$, T_{end} is the temperature used to control the total number of iterations).

The difference between SA, MMD and ICM can be appreciated looking at (3.34), specifically at the second case: if removed, one gets deterministic ICM procedure; if a is chosen once and for all at the start of the algorithm, one gets deterministic MMD; and finally, if is randomly chosen on every iteration, the result is stochastic SA [39].

Therefore, the algorithm steps can be summarized as:

Learning Steps: Potential energy functions of each class are learned on training areas selected by the user. Parameters of predefined distribution models are based on the amplitude of pre-classified pixels;

Classification Step: Once the potential energy functions are known, they are integrated in the Markov field. We make use of MMD algorithm to accomplish the classification process.

3.3.3 Results and Analysis

A NASA/JPL AIRSAR L-band HH polarized image of San Francisco, shows in Fig. 3.12a, is used. Figure 3.12b is the ground truth map manually segmented. The size of the image is 400 × 500 pixels. Three classes are in the scene, which are building, forest and sea, numbered with 1, 2, and 3, respectively. The size of each class of training data is 70 × 40 pixels.

By applying the real SAR data denoting different terrain categories, the clutter is analyzed with amplitude histogram fitting, compared with Nakagami distribution.

Nakagami distribution, whose intensity form is called gamma distribution, can accurately model the homogeneous clutter regions. Figure 3.13 shows the histograms of sample data used within each class and the estimated amplitude using two different kinds of distributions, respectively. In order to emphasize the details, normalization was performed on the original amplitude of the data.

From Fig. 3.13, it can be concluded that both G^0 and Nakagami give good description of the region of forest and sea, which are of the category of homogeneous clutter regions, with G^0 the better choice. However, it fails for Nakagami to model the data of building, the heterogeneous clutter region, whilst G^0 is still a good model. The modeling capability of G^0 distribution as a commonly used model and the efficiency of the parameter estimation method MoLC are verified.

After the process above, we obtain the PDFs of different kinds of land cover typologies. Based on this, we perform the Markovian classification framework on the whole image. To optimize the labeling problem, the initial label configuration is chosen by three approaches: random initialization, initialization by K-means algorithm and K Nearest Neighbor algorithm. The predefined parameters are: $a = 0.025$, $\beta = 0.4$, $T_0 = 4.0$, $T_{end} = 0.5$, ensuring that each site is visited about 250 times.

For comparison, classification maps using the proposed method, ICM algorithm and SA algorithm are shown in Fig. 3.14. In the figure, each row depicts the classification results of different optimize techniques with one of the three initial configurations above. The classification accuracy is shown in Table I, because of the senseless of the accuracy evaluation using ICM algorithm initialized randomly, they are marked "×" instead. Overall accuracy and Kappa coefficients [40] are suggested to evaluate the accuracy of the classification results.

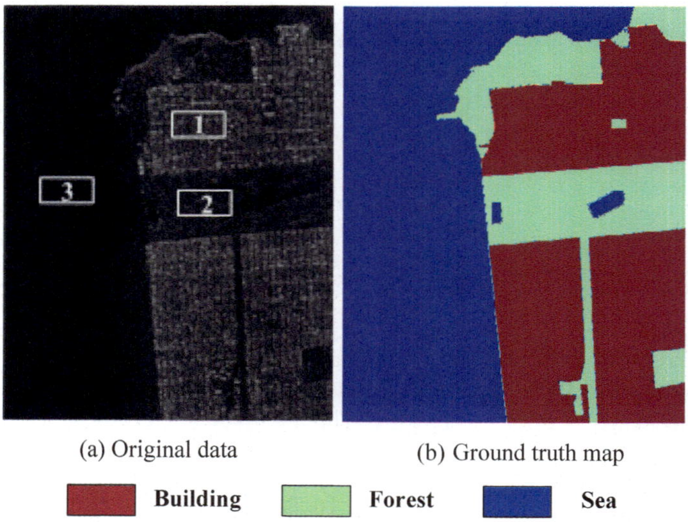

Fig. 3.12 Image data of San Francisco

Fig. 3.13 Histograms and estimated amplitudes

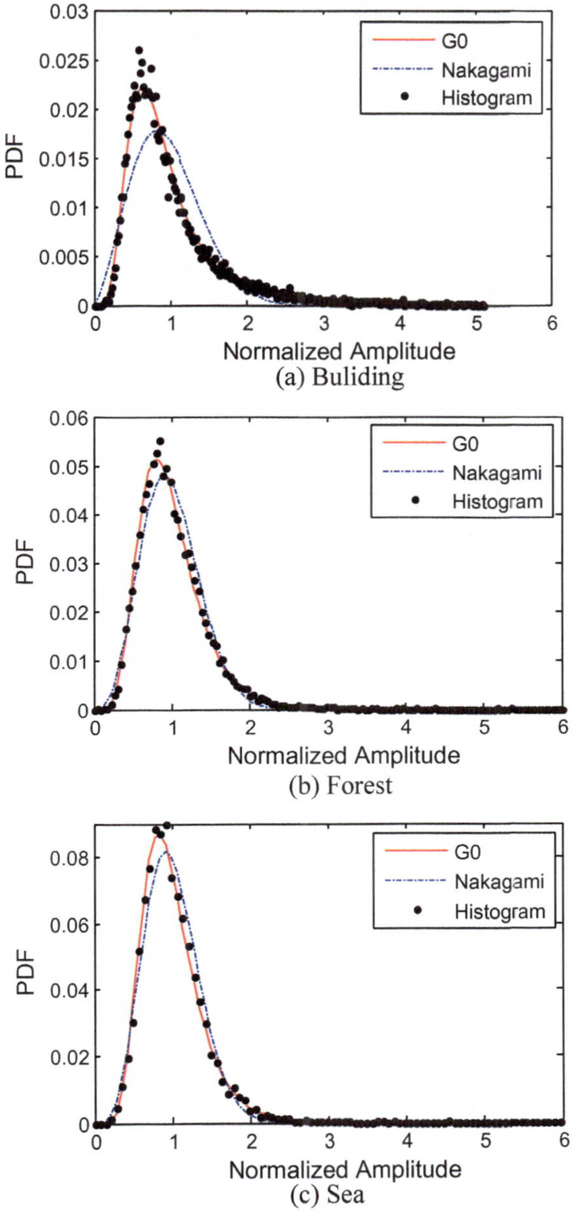

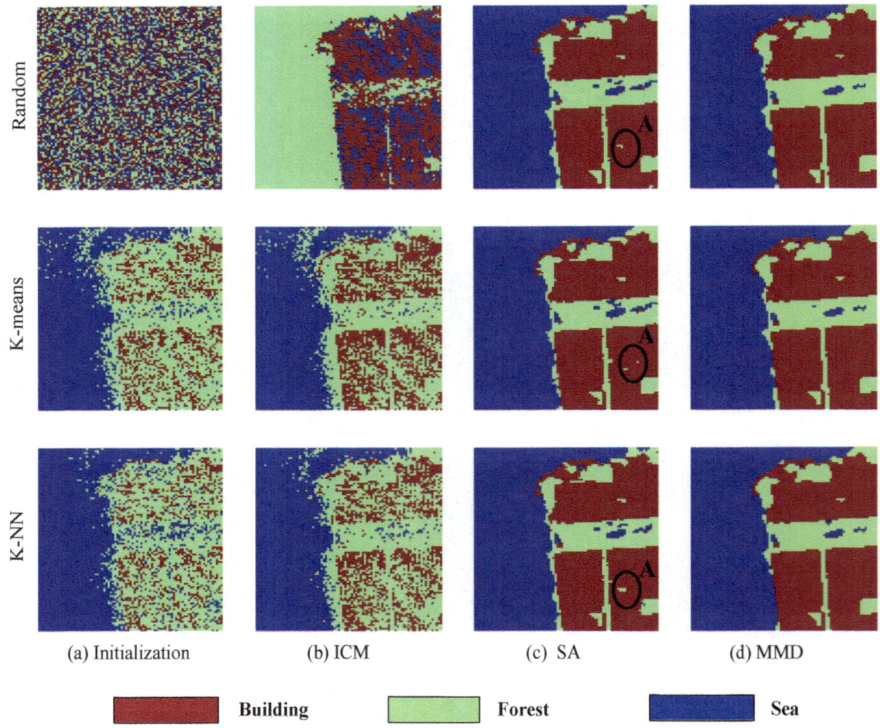

Fig. 3.14 Classification results using different optimize algorithm with different initializations

From the Fig. 3.14 and Table 3.6, it can be concluded that the classification performance using MMD algorithm is better, with the overall accuracy about 96% and Kappa coefficient beyond 0.93 in despite of different initial configurations. The classification map is smoother and there are less isolated regions and pixels, with better connectivity and smoother border. MMD and SA are both of better classification performance than ICM, mostly because MMD and SA can find the global minimum, on the contrary, ICM can only finds the local minimum. It is shown in Fig. 3.14b that the classification accuracy using ICM algorithm varies with the initial configuration, especially the configuration randomly chosen.

For the comparison between MMD and SA, with the same initialization and number of iteration, MMD can more quickly converge to the optimum result, which can be seen from not only the better classification results but also the edge preserving capability. The better performance of MMD in eliminating isolated pixels is also exhibited compared with SA, in area A, in Fig. 3.14c. As a whole, the classification result using MMD optimization algorithm is in good agreement with the ground truth map.

Table 3.6 The accuracy of classification using different optimize algorithms

	Randomly initialization			K-means			K-NN		
	ICM	SA	MMD	ICM	SA	MMD	ICM	SA	MMD
Building	×	99.3%	99.5%	33.1%	99.3%	99.5%	20.4%	99.0%	99.4%
Forest	×	82.0%	82.7%	95.4%	82.3%	83.4%	90.0%	81.2%	82.0%
Sea	×	97.5%	97.4%	88.6%	97.4%	98.1%	96.0%	97.4%	98.2%
Overall accuracy	×	95.5%	95.7%	69.4%	95.5%	96.1%	67.2%	95.2%	95.9%
Kappa coefficient	×	0.927	0.930	0.552	0.928	0.937	0.518	0.923	0.933

3.4 Conclusion

We proposed a CFAR algorithm using a versatile GΓD as the underlying distribution for background. An analytical expression for the adaptive threshold of the algorithm is derived. The experimental results on the actual high-resolution SAR image show that the proposed algorithm has an excellent CFAR maintaining performance compared with the Gaussian, Weibull, K_A and G_A^0 distributed algorithms. Besides, it is significantly more efficient than the K_A and G_A^0 distributed algorithm. Therefore, it is attractive to apply the proposed algorithm in the practical application especially for the design of online target detection system.

For the future work, it is of great interest to extend the algorithm for computing the inverse incomplete gamma distribution to the general platform not only the Matlab.

Aiming at ship detection in SAR images, this chapter also proposes a Parzen Window Kernel Based CFAR Algorithm. The idea is firstly using a non-parametric methods based on Parzen window kernel to estimate the PDF of SAR image data with high estimation accuracy. Then, a numerical solution of the threshold is derived. The analysis of the detection performance over the typical real SAR images confirms the effectiveness of the proposed algorithm.

Further investigations on whether some simpler techniques such as look up tables can be employed need to be done, which may greatly reduce the time consumption of detection.

Additionally, Many isolated pixels are produced inevitably in classification maps using pixel-based methods due to speckle noises in SAR images, which make it difficult for getting good classification accuracy and understanding the images. In the method proposed in this chapter, by mixing G^0 distribution and markovian classification together, the statistical property of clutter and the spatial context information between adjacent pixels are exploited. MMD algorithm, suggested in the optimization process, is proven an efficient choice and a compromise solution between determined ICM algorithm and SA algorithm. In addition, the classification process is performed without operation like image enhancement and the classification accuracy is independence of the initial configuration.

References

1. L.M. Novak, G.J. Owirka, C.M. Netishen, Performance of a high-resolution polarimetric SAR automatic target recognition system. Linc. Lab. J. **6**(1), 11–24 (1993)
2. L.M. Novak, G.J. Owirka, W.S. Brower, A.L. Weaver, The automatic target-recognition system in SAIP. Linc. Lab. J. **10**(2), 187–202 (1997)
3. W.C. Phillips, SAR image understanding: high speed target detection and site model based exploitation, Ph.D. dissertation, University of Maryland at College Park, 1998
4. S. Kuttikkad, R. Chellappa, Non-Gaussian CFAR techniques for target detection in high resolution SAR images. Proc. ICIP **1**, 910–914 (1994)
5. M. E. Smith, P. K. Varshney, *in Proceedings of IEEE National Radar Conference*. VI-CFAR: a novel CFAR algorithm based on data variability (*Syracuse,* NY, 1997), pp. 263–268
6. G. Gao, L. Liu, L. Zhao, G. Shi, G. Kuang, An adaptive and fast CFAR a Algorithm based on automatic censoring for target detection in high-resolution SAR image. IEEE Trans. Geosci. Remote Sens. **47**(6), 1685–1697 (2009)
7. S. Erfanian, V.T. Vakili, Introducing excision switching-CFAR in K distributed sea clutter. Signal Process. **89**, 1023–1031 (2009)
8. O.H. Bustos, M.M. Lucini, A.C. Frery, M-estimators of roughness and scale for *G0 A*-modelled SAR imagery. EURASIP J. Appl. Sig. Process. **2002**(1), 105–114 (2002)
9. H. Allende, A.C. Frery, J. Galbiati, L. Pizarro, M-estimators with asymmetric influence functions: the G_A^0 distribution case. IEEE Trans. Geosci. Remote Sens. **76**(11), 941–956 (2006)
10. R. Ravid, N. Levanon, Maximum-likehood CFAR for Weibull background. IEE Proc. F-Radar Sig. Process. **139**(3), 256–264 (1992)
11. C. Wang, M. Liao, X. Li, Ship detection in SAR image based on the alpha-stable distribution. Sensors 4948–4960 (2008)
12. M. Liao, C. Wang, Y. Wang, L. Jiang, Using SAR images to detect ships from sea clutter. IEEE Geosci. Remote Sens. Lett. **5**(2), 194–198 (2008)
13. H.C. Li, W. Hong, Y.R. Wu, P.Z. Fan, On the empirical-statistical modeling of SAR images with generalized gamma distribution. IEEE J. Sel. Top. Sig. Process. **5**(3), 386–397 (2011)
14. E.W. Stacy, A generalization of the gamma distribution. Ann. Math. Statist. **33**(3), 1187–1192 (1962)
15. J.M. Nicolas, Stian Normann Anfinsen (translator), "Introduction to second kind statistic: application of log-moments and log-cumulants to SAR image law analysis". Trait. Signal **19**(3), 139–167 (2002)
16. M. Abramowitz, L.A. Stegun, *Handbook of Mathematical Functions* (Dover, New York, 1972)
17. Harry, U.: Hansen's method applied to the inversion of the incomplete gamma function, with applications. IEEE Trans. Aerosp. Electro. Syst. **21**(5), 728–731 (1985)
18. M. Dohler, M. Arndt, Inverse incomplete gamma function and its application. Electron. Lett. **42**(1), 46–47 (2006)
19. A.C. Frery, H.J. Muller, C.C.F. Yanasse, S.J.S. Sant'Anna, A model for extremely heterogeneous clutter. IEEE Trans. Geosci. Remote Sens. **35**(3), 648–659 (1997)
20. M.D. DeVore, J.A. O'Sullivan, Quantitative statistical assessment of conditional models for synthetic aperture radar. IEEE Trans. Image Process. **13**(2), 113–125 (2004)
21. M. Tello, C. López-Martínez, J.J. Mallorqui, A novel algorithm for ship detection in SAR imagery based on the wavelet transform. IEEE Geosci. Remote Sens. Lett. **2**(2), 201–205 (2005)
22. K. Ouchi, S. Tamaki, H. Yaguchi, M. Iehara, Ship detection based on coherence images derived from cross correlation of multilook SAR images. IEEE Geosci. Remote Sens. Lett. **1**(3), 184–187 (2004)
23. R.A. English, S.J. Rawlinson, N.M. Sandirasegaram, Development of an ATR workbench for SAR imagery. De Defense R&D, Ottawa, ON, Canada, Technical Report, DRDC Ottawa, TR2002-155, 2002
24. P.W. Vachon, Validation of ship detection by the RADARSAT synthetic aperture radar and the ocean monitoring workstation. Can. J. Remote. Sens. **26**(3), 200–212 (2000)

25. M.P. Wand, M.C. Jones, *Kernel Smoothing* (Chapman & Hall, 1995)
26. M. Silveira, S. Heleno, in *IEEE International Conference* on *Image Processing (ICIP)*. Classification of water region in SAR images using level sets and non-parametric density estimation (2009), pp. 1685–1688
27. J. Sun, *Modern Pattern Recognition* (Publishing House of National University of Defense Techonlogy, Changsha, 2002)
28. S. Erfanian, V.T. Vakili, Introducing excision switching-CFAR in K distributed sea clutter. Sig. Process. **89**(6), 1023–1031 (2009)
29. T.M. Cover, J.A. Thomas, *Elements of Information Theory* (Wiley Interscience, New York, 1991)
30. M.D. DeVore, J.A. O'Sullivan, Quantitative statistical assessment of conditional models for synthetic aperture radar. IEEE Trans. Image Processing **13**(2), 113–125 (2004)
31. C. Oliver, S. Quegan, *Understanding Synthetic Aperture Radar Images* (SciTech Publishing, Raleigh, 2004)
32. G. Moser et al., SAR amplitude probability density function estimation based on a generalized Gaussian model. IEEE Trans. Image Process. **15** (2006)
33. G. Moser et al., *in IS&TSPIE Electronic Imaging*. High resolution SAR-image classification by Markov random fields and finite mixtures (2010)
34. A.C. Frery et al., A model for extremely heterogeneous clutter. IEEE Trans. Geosci. Remote Sens. **35**, 648–659 (1997)
35. J.-M. Nicolas, A. Maruani, in *EUSIPCO,* Tampere, Finland. Lower-order statistics: a new approach for probability density functions defined on R^+ (2000)
36. S. Geman, D. Geman, Stochastic relaxation, Gibbs distributions, and the Bayesian restoration of Images. IEEE Trans. Pattern Anal. Mach. Intell. **6**, 721–741 (1984)
37. D.P. Kottk et al., in *SPIE*, Orland, Florida. Design for HMM-based SAR ATR (1998)
38. Z. Kato et al., in *IEEE International Conference on Acoustics, Speech, and Signal Processing*. Satellite image classification using a modified metropolis dynamics (1992), pp. 573–576
39. V. Krylov, J. Zerubia, *High Resolution SAR Image Classification* (INRIA, Paris, 2010)
40. G.M. Foody, Status of landcover classification accuracy assessment. Remote Sens. Environ. **58**, 1459–1460 (1992)

Chapter 4
Statistical Modeling of Multi-channel SAR Images

4.1 Introduction

Currently, with the advancement of sophisticated SAR imaging modes, such as multi-temporal interferometry and polarimetry, the return backscatter results are presented in two or more channels [1, 2]. The SAR interferogram, achieved by multiplying the first image by the complex conjugate of the second one [3, 4], has become an important tool for multiple-channel SAR image interpretation. As a prominent example of interferogram applications, ground moving target indication (GMTI) has also received a great deal of attention and been intensively studied, e.g., [5–9].

Ground moving target identification (GMTI) using SAR has been a growing interest over the last couple of decades in many applications, such as military surveillance and reconnaissance of ground vehicles, and civilian ship monitoring of harbor [1–3]. The recent works [3, 4] reported in this field show that the multilook interferogram is an important tool for detecting moving targets. However, precise knowledge of the interferogram's phase and magnitude statistics, i.e., the joint probability density function (PDF), is a major issue currently under study in the development of statistically based detector tests for distinguishing the moving targets from clutter [3–5].

Some investigations for statistical modeling of multilook SAR interferogram have been presented in the past, e.g., [3–5]. Lee et al. [5] firstly proposed the joint distribution of interferometric magnitude and phase with the condition of a constant radar cross section (RCS) background based on the complex Wishart distribution, presented by Goodman [6]. In the analysis of lot of literatures, it is shown [3–5] that the PDF is valid for modeling homogeneous areas, whereas it also tends to deviate strongly in most cases whose scenes contain heterogeneous or extremely heterogeneous regions.

Additionally, as the phase statistic is highly invariant against changes of the clutter type [3], the marginal PDFs of interferometric magnitude for heterogeneous and extremely heterogeneous regions are derived by Gierull [4]. Meanwhile, an original joint PDF of interferometric magnitude and phase for heterogeneous clutter is also

G. Gao, *Characterization of SAR Clutter and Its Applications to Land and Ocean Observations*, https://doi.org/10.1007/978-981-13-1020-1_4

given in afore-mentioned literature. Unfortunately, the joint PDFs of interferometric magnitude and phase for extremely heterogeneous regions are still a hard task by means of combining the marginal PDFs of magnitude and the ones of phase owing to that magnitude and phase are not statistically independent [3].

Meanwhile, since the theoretic foundation of the interferometric magnitude modeling is the complex Wishart distribution, some studies have been proposed based on this distribution in the past several years, such as [3, 4, 7–11]. Lee et al. [3] firstly proposed the interferometric magnitude distribution with the condition of a RCS background based on the complex Wishart distribution. Actually, the RCS of a homogeneous region (e.g., the agricultural areas) in either low-resolution or high-resolution SAR images can be expected to be a constant [12]. However, most scenes contain in-homogeneous regions with RCS fluctuations. Therefore, the magnitude's probability density function (PDF) tends to deviate strongly in most cases. In practice, the product model [12, 13] has been widely and successfully used in the studies of the statistics of the single-channel image magnitude. Under this consideration, the interferometric magnitude can also be regarded as the product between an underlying RCS component with an uncorrelated multiplicative speckle noise component, which is similar with the polarimetric SAR images [14, 15].

In this chapter, our objective is to present a novel joint distribution of interferometric magnitude and phase for extremely heterogeneous clutter. We test the performance of the proposed distribution utilizing a representative dual-channel SAR image of urban area described as an extremely heterogeneous region.

Recently, Frery et al. [14–16] propose the reciprocal of the square root of a Gamma and the square root of a generalized inverse Gaussian to describe the signal components (RCS fluctuations) of SAR image magnitude for heterogeneous and extremely heterogeneous terrains. Motivated by this idea, based on famous complex Wishart distribution, this chapter utilizes the product model to deduce analytically the distribution models of interferogram's magnitude under different environments, these distributions are simply referred to as the $\Gamma_{\mathcal{I}n}$ distribution, the $\mathcal{K}_{\mathcal{I}n}$ distribution and the $\mathcal{G}^0_{\mathcal{I}n}$ distribution (corresponded to homogeneous, heterogeneous and extremely heterogeneous regions respectively). The presented distribution models are the successful generalization from the statistical model family of single-channel SAR images to the field of multi-channel SAR images. On this basis, using the second-kind statistics theory [17–19], i.e., MoLC, we derive the corresponding parameter estimators, as $\Gamma_{\mathcal{I}n}\_MoLC$, $\mathcal{K}_{\mathcal{I}n}\_MoLC$ and $\mathcal{G}^0_{\mathcal{I}n}\_MoLC$. These estimators are capable of obtaining the iterative results accurately.

4.2 Normalized Interferogram

According to the central limit theorem, when the RCS of clutter background is constant, the in-phase and quadrature components of speckle are independent and identically distributed. Both of them obey the zero-mean complex Gaussian distribution

[12], which meets the condition of the complex Wishart distribution. As the average of several independent samples, the n-look sample covariance matrix is given as [7]

$$\hat{\boldsymbol{R}} = \frac{1}{n}\sum_{k=1}^{n} \boldsymbol{Z}(k)\boldsymbol{Z}(k)^{\mathrm{H}} = \frac{1}{n}\sum_{k=1}^{n}\begin{bmatrix} |z_1(k)|^2 & z_1(k)z_2(k)^* \\ z_1(k)^* z_2(k) & |z_2(k)|^2 \end{bmatrix} \tag{4.1}$$

where n represents number of looks, $\boldsymbol{Z}(k) = [z_1(k), z_2(k)]^T$ is the kth single-look image, the superscript * represents complex conjugate and H means the complex conjugate transpose. The off-diagonal elements $(1/n)\sum_{k=1}^{n} z_1(k)z_2(k)^*$ indicate the complex n-look interferogram.

According to the literature [8], random matrix $\boldsymbol{B} = n\hat{\boldsymbol{R}}$ obeys the complex Wishart distribution

$$p_{\boldsymbol{B}}(\boldsymbol{B}) = \frac{|\boldsymbol{B}|^{n-2}\exp\left[-tr\left(\boldsymbol{C}^{-1}\boldsymbol{B}\right)\right]}{K(n,2)|\boldsymbol{C}|^n} \tag{4.2}$$

where $K(n, 2) = \pi\Gamma(n)\Gamma(n-1)$, the underlying covariance matrix $\boldsymbol{C}$ is

$$\boldsymbol{C} = \mathrm{E}\left[\boldsymbol{Z}\boldsymbol{Z}^{\mathrm{H}}\right] = \begin{bmatrix} C_{11} & \sqrt{C_{11}C_{22}}\rho e^{j\theta} \\ \sqrt{C_{11}C_{22}}\rho e^{-j\theta} & C_{22} \end{bmatrix} \tag{4.3}$$

where $\rho e^{j\theta}$ is the complex coefficient exported by the two channel. The magnitude of the complex correlation coefficient is represented by ρ, simply denotes as coherence magnitude in this chapter. Due to that the factors, such as thermal noise of receiver and speckle fluctuation noise, generate decorrelation, ρ is normally 0.95–0.99 in ground scenes. Besides, θ only relates to the imaging geometry and scene altitude, therefore we often suppose θ to be zero on ground scene.

In (4.2), the 4 variables are processed to be normalized, then integrating the main diagonal elements to obtain the joint distribution of the normalized interferogram's magnitude ξ and the multi-look phase ψ:

$$\begin{aligned} p_{\xi,\psi}(\xi,\psi) &= \frac{2n^{n+1}\xi^n}{\pi\Gamma(n)\left(1-\rho^2\right)}\exp\left(\frac{2n\rho\xi\,\cos(\psi-\theta)}{1-\rho^2}\right)K_{n-1}\left(\frac{2n\xi}{1-\rho^2}\right), \\ n, \xi, \rho &> 0 \quad \text{and} \quad -\pi < \psi < \pi \end{aligned} \tag{4.4}$$

where $K_{n-1}(\cdot)$ is the second type modified Bessel function with order $(n-1)$. $\Gamma(\cdot)$ indicates the Gamma function. Normalized magnitude ξ of interferogram is defined as

$$\xi = \frac{\left|(1/n)\sum_{k=1}^{n} z_1(k)z_2(k)^*\right|}{\sqrt{E\left(|z_1|^2\right)E\left(|z_2|^2\right)}} = \frac{\left|(1/n)\sum_{k=1}^{n} z_1(k)z_2(k)^*\right|}{\sqrt{C_{11}C_{22}}} \tag{4.5}$$

4.3 The Joint Distribution

4.3.1 The Known Joint Distribution for Heterogeneous Regions

Considering a n-look interferogram I_n, it is the average of n single-look interferograms. Assuming the energy of two channels is identical, it's well known that I_n can be modeled by the multiplicative model as [4]

$$I_n = A^2 \xi e^{j\psi} \tag{4.6}$$

where A represents the backscattering RCS magnitude of each channel.

As analyzed by Frery et al. [7], the random variable A obeys a reciprocal of the square root of a Gamma distribution to characterize highly heterogeneous situation, i.e., $A \sim \Gamma^{-1/2}(-\alpha, \gamma)$. Supposing $W = A^2$, the PDF of W is given by

$$p_W(w) = \frac{\gamma^{-\alpha}}{\Gamma(-\alpha)} w^{\alpha-1} \exp\left(-\frac{\gamma}{w}\right), \quad -\alpha, \gamma > 0 \tag{4.7}$$

where $\alpha(-\alpha \in (0, \infty))$ is a shape parameter, which essentially reflects the degrees of homogeneity for processed areas. γ is a scale parameter related to the mean energy of processed areas.

Therefore, the modified interferometric magnitude η of heterogeneous clutter is given as $\eta = A^2\xi := W\xi$ and the joint distribution of η and ψ can be expressed by

$$p_{\eta,\psi}(\eta, \psi) = \int_0^\infty p_W(w) p(\eta|w, \psi) \mathrm{d}w \tag{4.8}$$

Applying (4.4), we get the right hand-side of the integral shown in (4.8) as

$$\begin{aligned} p(\eta|w, \psi) &= \frac{1}{w} p\left(\frac{\eta}{w}, \psi\right) \\ &= \frac{2n^{n+1}\eta^n}{\pi\Gamma(n)\left(1-\rho^2\right)} \left(\frac{1}{w}\right)^{n+1} \exp\left(\frac{2n\eta\rho\cos\psi}{1-\rho^2} \times \frac{1}{w}\right) K_{n-1}\left(\frac{2n\eta}{1-\rho^2} \times \frac{1}{w}\right) \end{aligned} \tag{4.9}$$

Combining (4.7) and (4.9) by (4.8), and utilizing the integral formula $\int_0^\infty x^{\mu-1} e^{-ax} K_\nu(bx) \mathrm{d}x = \frac{\sqrt{\pi}(2b)^\nu}{(a+b)^{\mu+\nu}} \cdot \frac{\Gamma(\mu+\nu)\Gamma(\mu-\nu)}{\Gamma(\mu+1/2)} {}_2F_1\left(\mu+\nu, \nu+1/2; \mu+1/2; \frac{a-b}{a+b}\right)$ [8], the joint distribution of magnitude and phase in the heterogeneous clutter is finally derived as [4]

$$
\begin{aligned}
p_{\eta,\psi}(\eta,\psi) = & \frac{(2n)^{2n}\gamma^{-\alpha}}{2\sqrt{\pi}\left(1-\rho^2\right)^{-(n-\alpha)}} \cdot \frac{\Gamma(2n-\alpha)\Gamma(-\alpha+2)}{\Gamma(n)\Gamma(-\alpha)\Gamma\left(n-\alpha+3/2\right)} \cdot \eta^{2n-1} \\
& \cdot \left[f_1(\eta,\psi)\right]^{-(2n-\alpha)} \cdot {}_2F_1\left(2n-\alpha, n-\frac{1}{2}; n-\alpha+\frac{3}{2}; f_2(\eta,\psi)\right)
\end{aligned}
\tag{4.10}
$$

where ${}_2F_1$ is the Gauss hypergeometric function and

$$
\begin{cases}
f_1(\eta,\psi) = \left(1-\rho^2\right)\gamma + 2n\eta(1-\rho\cos\psi) \\
f_2(\eta,\psi) = \dfrac{\left(1-\rho^2\right)\gamma - 2n\eta(1+\rho\cos\psi)}{\left(1-\rho^2\right)\gamma + 2n\eta(1-\rho\cos\psi)}
\end{cases}
\tag{4.11}
$$

4.4 The Proposed Distribution for Interferogram's Magnitude of Homogenous Clutter

4.4.1 The $\Gamma_{\mathcal{I}n}$ Distribution for Homogeneous Clutter

To obtain probability distribution of interferogram's magnitude under homogeneous clutter background, the phase variable in (4.4) needs to be integrated. Via the following equation [21]:

$$
\int_0^{\pi} \exp(z\cos x)\mathrm{d}x = \pi I_0(z) \tag{4.12}
$$

The marginal PDF of normalized interferogram's magnitude can be obtained as [3]

$$
p_{\xi}(\xi) = \frac{4n^{n+1}\xi^n}{\Gamma(n)\left(1-\rho^2\right)} I_0\left(\frac{2n\rho\xi}{1-\rho^2}\right) K_{n-1}\left(\frac{2n\xi}{1-\rho^2}\right), \quad n,\xi,\rho > 0 \tag{4.13}
$$

where $I_0(\cdot)$ is the first type modified Bessel function of order zero. (4.13) is originally derived by Lee et al. [3].

The (4.13) has the following properties:

- The value of the first type modified Bessel function of order zero will tend to infinite rapidly [21] and $I_0(x) \to 1$ on the condition that $x \to 0^+$. See Fig. 4.1;
- Contrarily, the value of the second type modified Bessel function will tend to zero rapidly [21], see Fig. 4.2. Whereas $K_{n-1}(x) \approx \begin{cases} -\ln x, & n = 1 \\ \frac{\Gamma(n-1)}{2}\left(\frac{2}{x}\right)^{n-1}, & n > 1 \end{cases}$ when $x \to 0^+$;

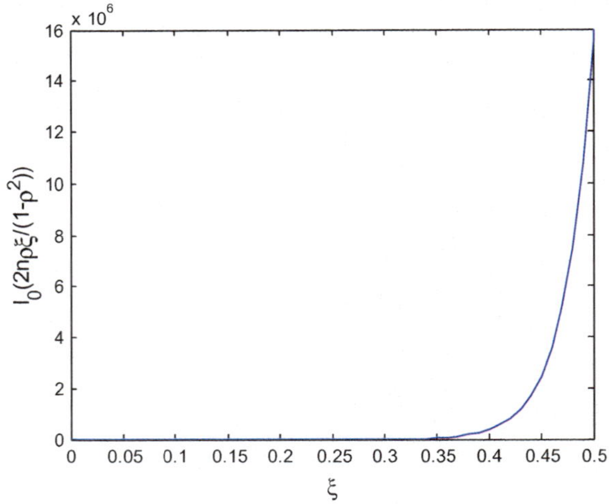

Fig. 4.1 Plot of the first type modified Bessel function of order zero with $n = 4$ and $\rho = 0.9$

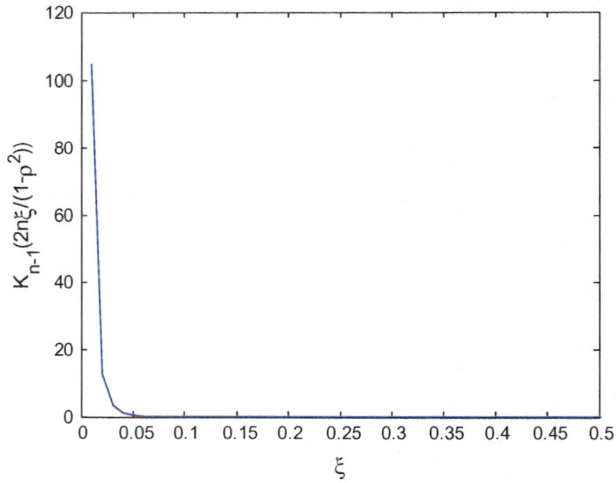

Fig. 4.2 Plot of the second type modified Bessel Function with $n = 4$ and $\rho = 0.9$

The two modified Bessel functions shown in (4.13) are generally evaluated by numerical methods. Owing to the complicated expressions of two types of modified Bessel function, the practical applications such as moving target detection by this distribution shown in (4.13) are limited, i.e., computation time, the analytical expression of detection threshold etc. It is a hard task to obtain the CFAR detection threshold by making an integral for (4.13).

Herein, our objective is to simplify the (4.13). Applying the asymptotic expansion formulas [21] of the first type and the second type modified Bessel functions, i.e.,

$$I_v(x) \cong \frac{\exp(x)}{\sqrt{2\pi x}} \sum_{k=0}^{K} \left[(-1)^k \frac{\Gamma\left(v + k + \frac{1}{2}\right)}{\Gamma(k+1)\Gamma\left(v + \frac{1}{2} - k\right)} \frac{1}{(2x)^k} \right]$$

$$K_\nu(x) \cong \frac{\sqrt{\pi}\,\exp(-x)}{\sqrt{2x}} \sum_{k=0}^{K}\left[\frac{\Gamma\left(\nu+k-\frac{1}{2}\right)}{\Gamma(k+1)\Gamma\left(\nu+\frac{1}{2}-k\right)}\frac{1}{(2x)^k}\right] \tag{4.14}$$

Then $I_0(c\rho)K_{n-1}(c)\ \left(c=\frac{2n\xi}{1-\rho^2}\right)$ in (4.13) can be denoted as

$$\begin{aligned} I_0(c\rho)K_{n-1}(c) &\cong \frac{\exp(-(1-\rho)c)}{2\sqrt{\rho}c}\left(\sum_{k=0}^{K}(-1)^k\frac{\Gamma\left(k+\frac{1}{2}\right)}{(2\rho c)^k\Gamma(k+1)\Gamma\left(\frac{1}{2}-k\right)}\right) \\ &\cdot\left(\sum_{k=0}^{K}\frac{\Gamma\left(n+k-\frac{1}{2}\right)}{(2c)^k\Gamma(k+1)\Gamma\left(n-\frac{1}{2}-k\right)}\right) \\ &= \frac{\exp(-(1-\rho)c)}{2\sqrt{\rho}} \\ &\sum_{k=0}^{K}\sum_{m=0}^{K}\frac{(-1)^k\Gamma\left(k+\frac{1}{2}\right)\Gamma^{-1}(k+1)\Gamma\left(n+m-\frac{1}{2}\right)}{2^{k+m}\rho^k c^{k+m+1}\Gamma\left(\frac{1}{2}-k\right)\Gamma(m+1)\Gamma\left(n-\frac{1}{2}-m\right)} \end{aligned} \tag{4.15}$$

Further, the PDF of ξ in (4.13) is approximated by

$$\begin{aligned} p_\xi(\xi) &\cong \frac{n^n}{\Gamma(n)}\xi^{n-1}\exp\left(-\frac{2n}{1+\rho}\xi\right) \\ &\cdot\sum_{k=0}^{K}\sum_{m=0}^{K}\left[\frac{(-1)^k\Gamma\left(k+\frac{1}{2}\right)\Gamma^{-1}(k+1)\Gamma\left(n+m-\frac{1}{2}\right)}{\Gamma\left(\frac{1}{2}-k\right)\Gamma(m+1)\Gamma\left(n-\frac{1}{2}-m\right)\rho^{k+1/2}}\cdot\left(\frac{1-\rho^2}{4n\xi}\right)^{k+m}\right] \end{aligned} \tag{4.16}$$

As some examples shown in Fig. 4.3, we found that the dual summing part of (4.16) has a very small fluctuation in the range of $\xi \in \mathbb{R}^+$, which can be approximated[1] by a constant value m_0. Utilizing $\int_0^\infty p(\xi)\mathrm{d}\xi = 1$, $m_0 = 2^n/(1+\rho)^n$. We have the PDF of the ξ as

$$p_\xi(\xi) = \frac{\beta n^n}{\Gamma(n)}(\beta\xi)^{n-1}\exp(-n\beta\xi), \quad \beta, n, \xi > 0 \tag{4.17}$$

where $\beta = 2/(1+\rho)$.

Making a Definition of $\tau = \sqrt{C_{11}C_{22}}\xi := \sigma\xi$, hence τ represents the interferogram's magnitude. The PDF of τ is given analytically by transforming (4.17) as $\Gamma_{\mathcal{I}n}(n,\beta,\sigma)$, i.e.,

$$p_\tau(\tau) = \frac{1}{\Gamma(n)}\left(\frac{n\beta}{\sigma}\right)^n\tau^{n-1}\exp\left(-\frac{n\beta}{\sigma}\tau\right), \quad \beta, n, \sigma, \tau > 0 \tag{4.18}$$

---

[1]Note that it is very hard to give the proof that the dual summing part of (4.25) is a const in theory. However, we also found that this dual sum is almost a const as shown in Fig. 4.3 with varying n and ρ.

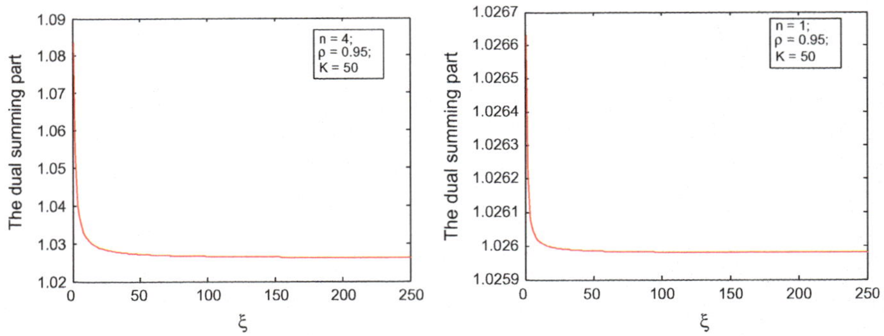

Fig. 4.3 The values of the dual summing part in (4.16) under different number of looks with $K = 50$ and $\rho = 0.95$

The (4.18) is the distribution of interferogram's magnitude of homogeneous clutter, hereafter simply as the $\Gamma_{\mathcal{I}n}$ distribution.

Once $\rho = 1$ (i.e., two channel outputs are completely correlated), $\Gamma_{\mathcal{I}n}$ is degenerated as

$$p_\tau(\tau) = \frac{1}{\Gamma(n)}\left(\frac{n}{\sigma}\right)^n \tau^{n-1} \exp\left(-\frac{n}{\sigma}\tau\right), \quad n, \sigma, \tau > 0 \tag{4.19}$$

The (4.19) is exactly the classical *Gamma* distribution $\Gamma(n, n/\sigma)$ [12, 16].

4.4.2 Parameter Estimators of $\Gamma_{\mathcal{I}n}$

(1) *Coherence magnitude estimation*

Normally, the estimation of coherence magnitude ρ is derived from the following equation [10]

$$\hat{\rho} = \left|\sum_{k=1}^{N} z_1(k) z_2(k)^*\right| \Bigg/ \sqrt{\sum_{k=1}^{N} |z_1(k)|^2 \sum_{k=1}^{N} |z_2(k)|^2} \tag{4.20}$$

where N is sample number, the PDF of estimation $\hat{\rho}$ is given as

$$f(\hat{\rho}) = 2(N-1)\left(1-\rho^2\right)^N \hat{\rho}\left(1-\hat{\rho}^2\right)^{N-2} {}_2F_1\left(N, N; 1; \rho^2\hat{\rho}^2\right) \tag{4.21}$$

where ${}_pF_q$ is the generalized hypergeometric function. In [10], the first moment of $\hat{\rho}$ is

$$\mathrm{E}(\hat{\rho}) = \frac{\Gamma(N)\Gamma(3/2)}{\Gamma(N+1/2)}\left(1-\rho^2\right)^N \cdot {}_3F_2\left(3/2, N, N; N+1/2, 1; \rho^2\right) \tag{4.22}$$

This classical estimation shown in (4.20) is biased [10]. Therefore, starting from the definition of normalized interferogram $\mathcal{I}_n = (1/n)\sum_{k=1}^{n} z_1(k)z_2(k)^* \Big/ \sqrt{\mathrm{E}(|z_1|^2)\mathrm{E}(|z_2|^2)}$, this chapter applies (4.4) to derive 1th order moment of $\mathcal{I}_n$ (see Appendix 4.1). The coherence magnitude estimation can be described exactly as

$$\hat{\rho} = |\mathrm{E}(\mathcal{I}_\mathrm{n})| \tag{4.23}$$

This estimation is unbiased as $\hat{\rho} = \mathrm{E}(\hat{\rho}) = |\mathrm{E}(\mathcal{I}_\mathrm{n})|$. After finishing the estimation of ρ, β is a constant in the $\Gamma_{\mathcal{I}n}$ distribution.

(2) $\Gamma_{\mathcal{I}n}$*_MoLC*

The second-kind first characteristic function and the second-kind second characteristic function of the $\Gamma_{\mathcal{I}n}$ distribution are:

$$\begin{cases} \phi_{\Gamma_{\mathcal{I}n}}(s) = \left(\frac{\sigma}{\beta n}\right)^{s-1} \frac{\Gamma(n+s-1)}{\Gamma(n)} \\ \zeta_{\Gamma_{\mathcal{I}n}}(s) = (s-1)\ln\left(\frac{\sigma}{\beta n}\right) + \ln\Gamma(n+s-1) - \ln\Gamma(n) \end{cases} \tag{4.24}$$

The $\Gamma_{\mathcal{I}n}$ distribution's log-cumulants are given by

$$\begin{cases} \tilde{c}_1 = -\ln\left(\beta n/\sigma\right) + \Psi(n) \\ \tilde{c}_k = \Psi(k-1, n), \quad k \geq 2 \end{cases} \tag{4.25}$$

where $\Psi(\cdot)$ represents the digamma function (i.e., the logarithmic derivative of the Gamma function), and $\Psi(k, \cdot)$ is the k th order polygamma function (i.e., the k th order derivative of the digamma function).

Given a sample set $\{x_i\}$, $i \in [1, N]$, The estimation expressions of parameters n and σ in the $\Gamma_{\mathcal{I}n}$ distribution can be derived as

$$\begin{cases} \Psi(\hat{n}) - \ln\left(\frac{\beta\hat{n}}{\hat{\sigma}}\right) = \frac{1}{N}\sum\limits_{i=1}^{N}[\ln(x_i)] \\ \Psi(1, \hat{n}) = \frac{1}{N}\sum\limits_{i=1}^{N}\left[\left(\ln(x_i) - \hat{\tilde{c}}_1\right)^2\right] \end{cases} \tag{4.26}$$

This equation shows the method of log-cumulants of the $\Gamma_{\mathcal{I}n}$ distribution, called $\Gamma_{\mathcal{I}n}$***_MoLC***.

4.5 Statistics of Multilook SAR Interferogram for In-homogeneous Clutter Based on $\Gamma_{\mathcal{I}n}$

4.5.1 *Extremely Heterogeneous Clutter*

For heterogeneous terrain like forest, cultivated farmland, RCS of them have fluctuations. Meanwhile, for extremely heterogeneous clutter like the urban areas, the histograms show the heavy trail [22–25]. The performance of the $\Gamma_{\mathcal{I}n}$ distribution decreases and it generates large deviations in fitting these magnitude data. The product model has been testified that it is valid for statistical modeling of single-channel SAR magnitude. Its expression is as follows [12, 16],

$$Y_i = A_i X_i, \quad i = 1, 2 \quad \text{and} \quad X_i \sim \mathcal{N}^{\mathbb{C}}(0, 1) \tag{4.27}$$

where A_i represents the backscattering RCS magnitude component. $\mathcal{N}^{\mathbb{C}}(0, 1)$ denotes the complex normal distribution with expectation 0 and variance 1. X_i indicates speckle noise component, i is the independent receiving channel.

Assuming the energy of two channel is equal to each other, i.e., $A_i \equiv A$, the interferogram's magnitude could be expressed as

$$\Xi = A^2 \xi = W \xi \tag{4.28}$$

Frery et al. recently proposed a reciprocal of the square root of a Gamma distribution for the modulating random variable A of heterogeneous clutter [14–16], i.e., $A \sim \Gamma^{-1/2}(-\alpha, \gamma)$, It can be shown that the PDF of W is

$$p_W(w) = \frac{\gamma^{-\alpha}}{\Gamma(-\alpha)} w^{\alpha-1} \exp\left(-\frac{\gamma}{w}\right), \quad -\alpha, \gamma > 0 \tag{4.29}$$

where $\alpha(-\alpha \in (0, \infty))$ is a shape parameter, which essentially reflects the degrees of homogeneity for processed areas. γ is a scale parameter related to the mean energy of processed areas.

Herein, within the structure of product model, combining the $\Gamma_{\mathcal{I}n}$ distribution and (4.29), and using the equation $\int_0^{\infty} x^{m-1} \mathrm{e}^{-(a+1)x} \mathrm{d}x = \frac{\Gamma(m)}{(a+1)^m}$, $m > 0, a > -1$ [21], the PDF of Ξ can be derived as $\mathcal{G}^0_{\mathcal{I}n}(n, \alpha, \beta, \gamma)$:

$$p_\Xi(\Xi) = \frac{\beta n^n \gamma^{-\alpha} \Gamma(n-\alpha)}{\Gamma(n)\Gamma(-\alpha)} \cdot \frac{(\beta \Xi)^{n-1}}{(\gamma + n\beta\Xi)^{n-\alpha}}, \quad \beta, -\alpha, \gamma, n, \Xi > 0 \tag{4.30}$$

Here (4.30) is called as $\mathcal{G}^0_{\mathcal{I}n}$ distribution. This model enables the modeling of areas with varying degrees of homogeneity owing to the importing of the reciprocal of a square root of Gamma distribution.

Especially, when the two channels are completely correlated, $\mathcal{G}_{\mathcal{I}n}^0$ is degenerating to the well-known intensity distribution $\mathcal{G}_I^0(\alpha, \gamma, n)$ [16, 26, 27]:

$$p_\Xi(\Xi) = \frac{n^n \gamma^{-\alpha} \Gamma(n-\alpha)}{\Gamma(n)\Gamma(-\alpha)} \cdot \frac{\Xi^{n-1}}{(\gamma + n\Xi)^{n-\alpha}}, \quad -\alpha, \gamma, n, \Xi > 0 \tag{4.31}$$

4.5.2 Heterogeneous Clutter

For heterogeneous clutter in a single-channel SAR image, the $\mathcal{K}$ distribution is obtained when assuming a multiplicative noise model for SAR amplitude by expressing the RCS component as a square root of Gamma distribution [14–16, 12], i.e., $A \sim \Gamma^{1/2}(\alpha, \lambda)$. We here rewrite (4.28) as $\ell = A^2\xi = \upsilon\xi$, the corresponding PDF of υ is expressed as

$$p_\upsilon(\upsilon) = \frac{\lambda^\alpha}{\Gamma(\alpha)} \upsilon^{\alpha-1} \exp(-\lambda\upsilon), \quad \alpha, \lambda, \upsilon > 0 \tag{4.32}$$

Thus, the deriving process is similar as that of $\mathcal{G}_{\mathcal{I}n}^0$ distribution. We utilize the product model and combine the $\Gamma_{\mathcal{I}n}$ distribution and (4.32), thus resulting in the following distribution (named for convenience as $\mathcal{K}_{\mathcal{I}n}$) via $\int_0^\infty x^{\alpha-1} \exp\left(-\frac{\gamma}{x} - \lambda x\right) dx = 2\left(\frac{\gamma}{\lambda}\right)^{\alpha/2} K_\alpha\left(2\sqrt{\gamma\lambda}\right)$ for the magnitude of interferogram:

$$p_\ell(\ell) = \frac{2\lambda\beta n}{\Gamma(n)\Gamma(\alpha)} \cdot (\lambda\beta n\ell)^{(\alpha+n)/2-1} \cdot K_{\alpha-n}\left(2\sqrt{\lambda\beta n\ell}\right), \quad \beta, \alpha, \lambda, n, \ell > 0 \tag{4.33}$$

This distribution can fit a wide range of experimental data well because the square root of Gamma distribution is a well known model for RCS component in homogeneous and heterogeneous clutter. Especially, when the two channels are completely correlated, $\mathcal{K}_{\mathcal{I}n}$ is degenerating to the well-known $\mathcal{K}$ intensity distribution $\mathcal{K}_I(\alpha, \lambda, n)$ [16, 12]:

$$p_\ell(\ell) = \frac{2\lambda n}{\Gamma(n)\Gamma(\alpha)} \cdot (\lambda n\ell)^{(\alpha+n)/2-1} \cdot K_{\alpha-n}\left(2\sqrt{\lambda n\ell}\right), \quad \alpha, \lambda, n, \ell > 0 \tag{4.34}$$

4.5.3 Parameter Estimators of In-homogeneous Clutter Statistics

(1) $\mathcal{K}_{\mathcal{I}n}$_MoLC

The second-kind first characteristic function and the second-kind second characteristic function of the $\mathcal{K}_{\mathcal{I}n}$ distribution are

$$\begin{cases} \phi_{\mathcal{K}_{\mathcal{I}n}}(s) = (\lambda\beta n)^{-(s-1)} \dfrac{\Gamma(\alpha+s-1)\Gamma(n+s-1)}{\Gamma(\alpha)\Gamma(n)} \\ \zeta_{\mathcal{K}_{\mathcal{I}n}}(s) = -(s-1)\ln(\lambda\beta n) + \ln\Gamma(\alpha+s-1) \\ \qquad + \ln\Gamma(n+s-1) - \ln\Gamma(\alpha) - \ln\Gamma(n) \end{cases} \tag{4.35}$$

The estimation of parameters in the $\mathcal{K}_{\mathcal{I}n}$ distribution can be derived as

$$\begin{cases} \Psi(\hat{n}) + \Psi(\hat{\alpha}) - \ln\left(\hat{\lambda}\beta\hat{n}\right) = \frac{1}{N}\sum\limits_{i=1}^{N}[\ln(x_i)] \\ \Psi(1,\hat{n}) + \Psi(1,\hat{\alpha}) = \frac{1}{N}\sum\limits_{i=1}^{N}\left[\left(\ln(x_i) - \hat{\tilde{c}}_1\right)^2\right] \\ \Psi(2,\hat{n}) + \Psi(2,\hat{\alpha}) = \frac{1}{N}\sum\limits_{i=1}^{N}\left[\left(\ln(x_i) - \hat{\tilde{c}}_1\right)^3\right] \end{cases} \tag{4.36}$$

This estimator is called as $\mathcal{K}_{\mathcal{I}n}$_***MoLC***.

(2) $\mathcal{G}^0_{\mathcal{I}n}$_*MoLC*

The second-kind first characteristic function and the second-kind second characteristic function of the $\mathcal{G}^0_{\mathcal{I}n}$ distribution, i.e.,

$$\begin{cases} \phi_{\mathcal{G}^0_{\mathcal{I}n}}(s) = \left(\dfrac{\gamma}{\beta n}\right)^{s-1} \cdot \dfrac{\Gamma(n+s-1)\Gamma(-\alpha-(s-1))}{\Gamma(n)\Gamma(-\alpha)} \\ \zeta_{\mathcal{G}^0_{\mathcal{I}n}}(s) = (s-1)\ln\left(\dfrac{\gamma}{\beta n}\right) + \ln\Gamma(n+s-1) + \ln\Gamma(-\alpha-(s-1)) \\ \qquad - \ln\Gamma(n) - \ln\Gamma(-\alpha) \end{cases} \tag{4.37}$$

Calculating the derivative of $\zeta_{\mathcal{G}^0_{\mathcal{I}n}}(s)$ at $s = 1$, the log-cumulants of the $\mathcal{G}^0_{\mathcal{I}n}$ distribution is described by

$$\begin{cases} \tilde{c}_1 = \ln\left(\frac{\gamma}{\beta n}\right) + \Psi(n) - \Psi(-\alpha) \\ \tilde{c}_k = \Psi(k-1,n) + (-1)^k\Psi(k-1,-\alpha), \quad k \geq 2 \end{cases} \tag{4.38}$$

The expressions of estimating parameters α, γ, n in the $\mathcal{G}^0_{\mathcal{I}n}$ distribution by (4.38) and (4.26) are given as

$$\begin{cases} \ln(\hat{\gamma}/(\beta\hat{n})) + \Psi(\hat{n}) - \Psi(-\hat{\alpha}) = \frac{1}{N}\sum\limits_{i=1}^{N}[\ln(x_i)] \\ \Psi(1,\hat{n}) + \Psi(1,-\hat{\alpha}) = \frac{1}{N}\sum\limits_{i=1}^{N}\left[\left(\ln(x_i) - \hat{\tilde{c}}_1\right)^2\right] \\ \Psi(2,\hat{n}) - \Psi(2,-\hat{\alpha}) = \frac{1}{N}\sum\limits_{i=1}^{N}\left[\left(\ln(x_i) - \hat{\tilde{c}}_1\right)^3\right] \end{cases} \tag{4.39}$$

Similarly, the above equation is called as $\mathcal{G}^0_{\mathcal{I}n}$_***MoLC***.

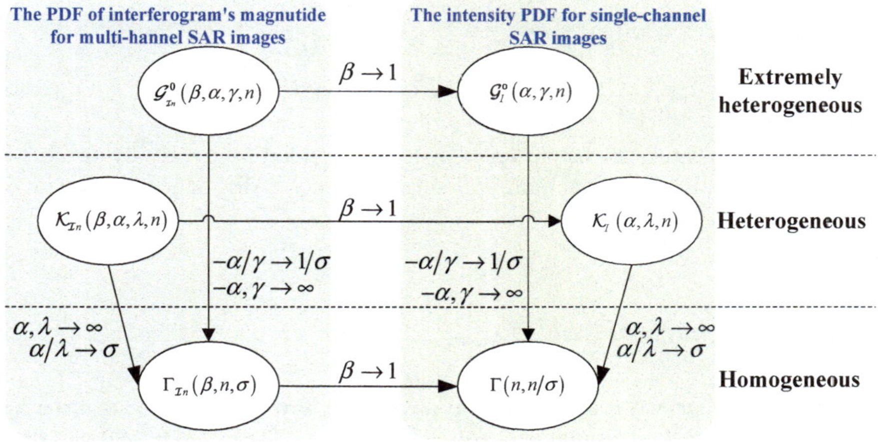

Fig. 4.4 The relationship of distributions

4.5.4 *Relationship Between Distributions*

Figure 4.4 shows the relationship of presented distributions mentioned-above. It is clear that these distributions have the following properties.

(1) *The interferogram's magnitude PDFs are "downward compatible"*. It can be proved that the $\mathcal{G}^0_{\mathcal{I}n}(\beta,\alpha,\gamma,n)$ converges in distribution to the $\Gamma_{\mathcal{I}n}(\beta,n,\sigma)$ when $-\alpha,\gamma\to\infty$ and $-\alpha/\gamma\to 1/\sigma$. The $\mathcal{K}_{\mathcal{I}n}(\beta,\alpha,\lambda,n)$ converges in distribution to the $\Gamma_{\mathcal{I}n}(\beta,n,\sigma)$ when $\alpha,\lambda\to\infty$ and $\alpha/\lambda\to\sigma$. Moreover, $\mathcal{G}^0_{\mathcal{I}n}$ encompasses the modeling abilities of $\mathcal{K}_{\mathcal{I}n}$ whilst extending them to enable the modeling of extremely heterogeneous data, because of the empirical evidence [16] describing the relationship of the square root of Gamma distribution and the reciprocal of a square root of Gamma distribution. The properties stated in Fig. 4.4 show that either homogeneous, heterogeneous, or extremely heterogeneous interferogram's magnitude statistics can be treated as the outcome of the $\mathcal{G}^0_{\mathcal{I}n}$ distribution.
(2) *The interferogram's magnitude PDFs are "channel compatible"*. Herein, the "channel compatible" means that the interferogram's magnitude PDF of multi-channel SAR images is compatible with intensity PDF of single channel SAR images. When the two channel outputs are completely correlated, the $\Gamma_{\mathcal{I}n}$, the $\mathcal{K}_{\mathcal{I}n}$, and the $\mathcal{G}^0_{\mathcal{I}n}$ distributions are respectively degenerated to the Γ, the $\mathcal{K}_I$, and the $\mathcal{G}^0_I$ distributions, see (4.19), (4.34), and (4.31). These three known distributions are extensively used in modeling single channel SAR intensity images [16, 20].

Additionally, from Fig. 4.4 and Eqs. (4.18), (4.19), (4.30), (4.31), (4.33) and (4.34), due to β can be regarded as a constant, we have the following results:

$$\begin{aligned}\Gamma_{\mathcal{I}n}(n,\beta,\sigma) &= \Gamma\left(n,\sigma/\beta\right)\\ \mathcal{K}_{\mathcal{I}n}(n,\alpha,\beta,\lambda) &= \mathcal{K}_I(n,\alpha,\lambda\beta)\\ \mathcal{G}^0_{\mathcal{I}n}(n,\alpha,\beta,\gamma) &= \mathcal{G}^0_I\left(n,\alpha,\gamma/\beta\right)\end{aligned} \tag{4.40}$$

This property indicates the presented distribution models successfully generalize the powerful statistical model family of single-channel SAR images to the field of multi-channel SAR images.

4.5.5 Experimental Analysis

In this section, our major objective is to investigate how the proposed distribution models perform well on the really measured InSAR data. The tested InSAR data used in this study were collected by NASA/JPL's airborne platform (known as AirSAR) operated in C band and standard dual-baseline ATI mode, with the spatial resolution 3.331 m × 3.898 m (range × azimuth). The name of this flight-line (site name) is usamacinta245-1, and the imaging time of ground scene is March 4, 2004. The original ground scene is very large and abundant terrain classes are included. The detailed imaging parameters corresponding to this scene are listed in Table 4.1.

For convenience of displaying, a local image of normalized interferometric magnitude from this scene is shown in Fig. 4.5 in dB format. The horizontal and vertical axes are the directions of azimuth and range, respectively. Four patches indicated by boxes in Fig. 4.5, numbered A–D, are the main areas of this investigation. The chosen patches consist of very different terrain, form homogeneous grassland area (Patch A) to heterogeneous area (Patch B) dominated by vegetation of trees, even extremely heterogeneous urban areas (Patches C and D), and hence, provide a wide variety of scattering properties.

Figures 4.6 show the fitting results of the proposed distributions for the four areas indicated in Fig. 4.5. As seen from Fig. 4.6, in order to reveal the details more clearly, the square root of the normalized interferometric magnitude is adopted to suppress the dynamic range. The parameter estimations of each distribution are all accomplished by the proposed estimators based on the MoLC, which are shown at Table 4.2.

In order to quantitatively assess the fitting result, we adopt KL (Kullback-Leibler) distance [28] as a similarity measurement. This measurement is the means of the global comparison of PDFs [28].

When the actual density equals to theoretical density, the KL distance D_{KL} is zero. Otherwise, D_{KL} is a positive value. The KL distance measurement reflects the overall similarity of the actual and theoretical densities. The smaller the value of KL distance measurement obtains, the higher similarity they have, which shows that fitting accuracy is better.

Table 4.1 The imaging parameters of the ground scene on usamacinta245-1 site

Latitude at start of scene in degrees	Longitude at start of scene in degrees	Latitude at end of scene in degrees	Longitude at end of scene in degrees	Width in Km	Length in Km	Chirp bandwidth	Radar altitude	Interferogram repeat time interval	Aircraft track angle in degrees	Number of lines in image	Number of samples per record
17.587143	−91.363557	17.371054	−91.710625	6.062	43.313	40 MHz	7997.018 m	0.00447 s	−114.796	11,113	1820

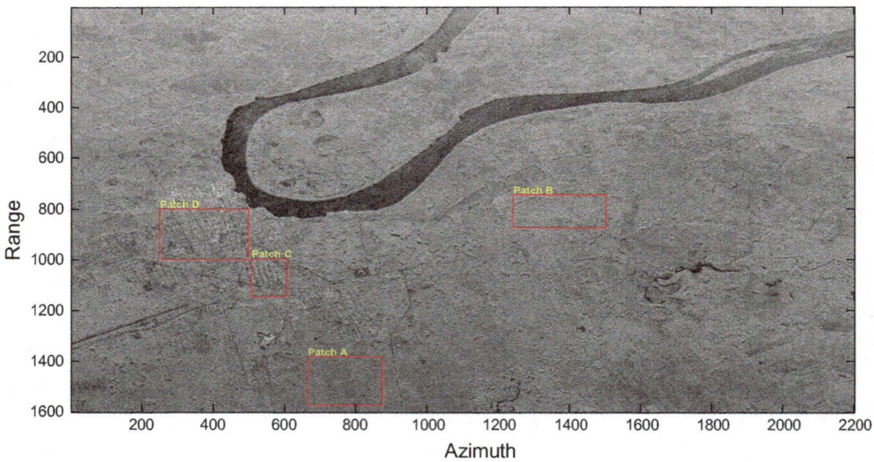

Fig. 4.5 The normalized interferometric magnitude image collected by the AirSAR on tested site

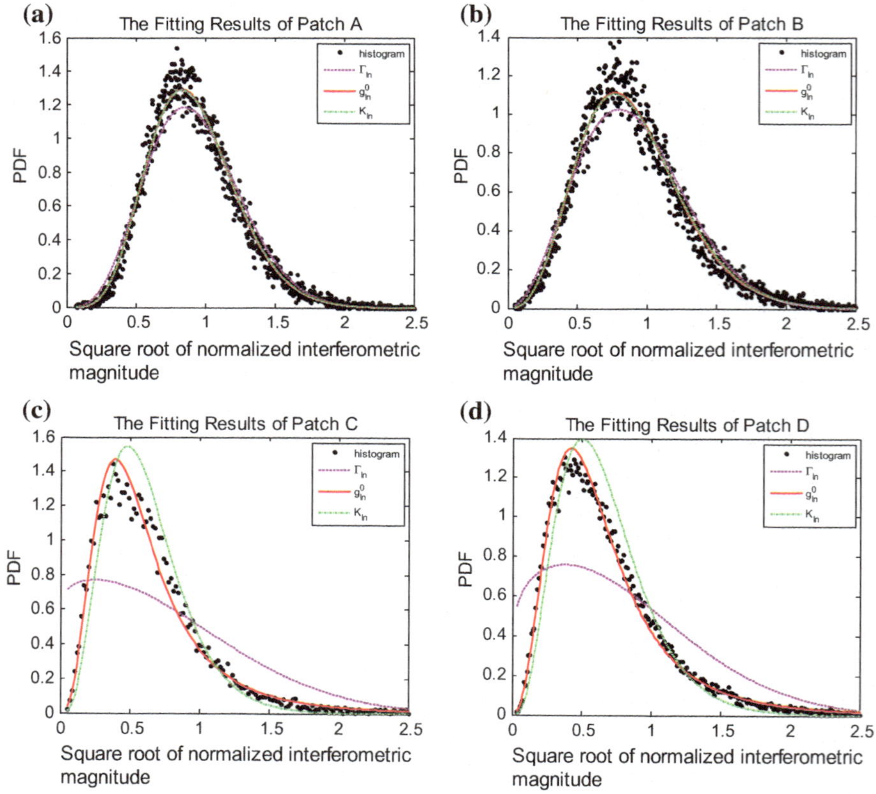

Fig. 4.6 Plots of interferometric magnitude histogram for the selected areas and of the estimated $\Gamma_{\mathcal{I}n}$, $\mathcal{K}_{\mathcal{I}n}$ and $\mathcal{G}^0_{\mathcal{I}n}$ PDFs: (**a–d**) are corresponded to Patches A–D, respectively

Table 4.2 Parameter estimations of noted clutter areas in Fig. 4.5

Patch No.	Terrain	$\hat{\rho}$	$\Gamma_{\mathcal{I}n}$ $(\hat{n})$	$\mathcal{K}_{\mathcal{I}n}$ $(\hat{\alpha}, \hat{\lambda}, \hat{n})$	$\mathcal{G}^0_{\mathcal{I}n}$ $(\hat{\alpha}, \hat{\gamma}, \hat{n})$
A	Grassland	0.9268	2.0054	(16.130, 16.591, 2.534)	(−18.275, 16.801, 2.501)
B	Tree	0.9412	1.4856	(9.125, 9.333, 1.847)	(−10.622, 9.418, 1.813)
C	Urban	0.9458	0.5321	(2.509, 5.353, 2.509)	(−1.516, 0.439, 1.779)
D	Urban	0.9423	0.5860	(2.334, 4.251, 2.334)	(−1.601, 0.549, 1.646)

Table 4.3 Statistical Characteristics of Different Patches in Fig. 4.5

Patch No.	Terrain	D_{KL}		
		$\Gamma_{\mathcal{I}n}$	$\mathcal{K}_{\mathcal{I}n}$	$\mathcal{G}^0_{\mathcal{I}n}$
A	Grassland	0.0454	0.0325	0.0311
B	Tree	0.0458	0.0403	0.0362
C	Urban	0.5390	0.2543	0.1083
D	Urban	0.3222	0.1804	0.0622

The KL values of the fitting results shown in Fig. 4.6 are compared in Table 4.3. It is evident that the following conclusions can be drawn, i.e.,

(1) For all four typical areas, the performance of $\Gamma_{\mathcal{I}n}$ is the relatively worst. However, the $\Gamma_{\mathcal{I}n}$ distribution tends to agree well with the areas with higher degree of homogeneity, which implies that $\Gamma_{\mathcal{I}n}$ is a proper statistical model for the homogeneous areas. Also, all the three densities tend to be alike for fitting areas with higher degree of homogeneity.

(2) With the increasing in-homogeneity degree of the data, the higher precision of fitting is obtained by using $\mathcal{K}_{\mathcal{I}n}$ than by using $\Gamma_{\mathcal{I}n}$. However, both $\mathcal{K}_{\mathcal{I}n}$ and $\Gamma_{\mathcal{I}n}$ give very poor statistical description of extremely heterogeneous urban areas, whilst $\mathcal{G}^0_{\mathcal{I}n}$ is a good model, whose KL value have a very small fluctuation in all.

(3) All the four areas with varying degrees of homogeneity are fitted well by $\mathcal{G}^0_{\mathcal{I}n}$. In another words, the $\mathcal{G}^0_{\mathcal{I}n}$ distribution shows greater capability of modeling different clutter data with a broad variety of scattering properties.

Appendix 4.1

The Estimation of Coherence Magnitude

The 1th order moment estimation of normalized interferogram $\mathcal{I}_n$ is

$$\begin{aligned} \mathrm{E}(\mathcal{I}_n) &= \int_0^{\infty}\int_{-\pi}^{\pi} \xi \mathrm{e}^{j\psi} \cdot p_{\xi,\psi}(\xi,\psi)\mathrm{d}\psi \mathrm{d}\xi \\ &= \int_0^{\infty} \frac{2n^{n+1}\xi^{n+1}}{\pi\Gamma(n)\left(1-\rho^2\right)} K_{n-1}\left(\frac{2n\xi}{1-\rho^2}\right)\left(\int_{-\pi}^{\pi} \mathrm{e}^{j\psi} \exp\left(\frac{2n\rho\xi\ \cos\psi}{1-\rho^2}\right)\mathrm{d}\psi\right)\mathrm{d}\xi \end{aligned} \tag{4.41}$$

Supposing $V = \int_{-\pi}^{\pi} \mathrm{e}^{j\psi} \exp\left(\frac{2n\rho\xi\ \cos\psi}{1-\rho^2}\right)\mathrm{d}\psi$, according to integral rules of odd and even functions, the definite integral of odd function is zero in a symmetry section, so it can be simplified as [21]

$$V = j2\pi J_1\left(-j\frac{2n\rho\xi}{1-\rho^2}\right) \tag{4.42}$$

where $J_1(\cdot)$ denotes the first type Bessel function of order 1. Plugging (4.42) to (4.41), then

$$\mathrm{E}(\mathcal{I}_n) = \frac{j4n^{n+1}}{\Gamma(n)\left(1-\rho^2\right)} \int_0^{\infty} \xi^{n+1} K_{n-1}\left(\frac{2n\xi}{1-\rho^2}\right) J_1\left(-j\frac{2n\rho\xi}{1-\rho^2}\right)\mathrm{d}\xi \tag{4.43}$$

From [21], (4.43) can be integrated as

$$\mathrm{E}(\mathcal{I}_n) = \rho\left(1-\rho^2\right)^{n+1} {}_2F_1\left(n+1, 2; 2; \rho^2\right) \tag{4.44}$$

where ${}_2F_1$ is the Gauss hypergeometric function. Using ${}_2F_1(n+1, m; m; z) = (1-z)^{-(n+1)}$, the 1st order moment estimation of $\mathcal{I}_n$ is finally given as

$$E(\mathcal{I}_n) = \rho \tag{4.45}$$

To make ρ be a positive real number, the estimation of ρ is expressed as

$$\rho = |E(\mathcal{I}_n)| \tag{4.46}$$

References

1. H. Steyskal, J.K. Schindler, P. Franchi, R.J. Mailloux, Pattern synthesis for TechSat21-A distributed space-based radar system. IEEE Antennas Propag. Mag. **45**(4), 19–25 (2003)
2. T. Wang, Z. Bao, Z. Zhang, J. Ding, Improving coherence of complex image pairs obtained by along-track bistatic SARs using range-azimuth prefiltering. IEEE Trans. Geosci. Remote Sens. **46**(1), 3–13 (2008)
3. J.S. Lee, K.W. Hoppel, S.A. Mango, A.R. Miller, Intensity and phase statistics of multilook polarimetric and interferometric SAR imagery. IEEE Trans. Geosci. Remote Sens. **32**(5), 1017–1028 (1994)
4. G. Ferraiuolo, G. Poggi, A Bayesian filtering technique for SAR interferometric phase fields. IEEE Trans. Image Process. **13**(10), 1368–1378 (2004)
5. D.C. Maori, J. Klare, A.R. Brenner, J.H.G. Ender, Wide-area traffic monitoring with the SAR/GMTI system PAMIR. IEEE Trans. Geosci. Remote Sens. **46**(10), 3019–3030 (2008)
6. E. Chapin, C.W. Chen, Along-track interferometry for ground moving target indication. IEEE Aerosp. Electron. Syst. Mag. **23**(6), 19–24 (2008)
7. C.H. Gierull, Statistical analysis of multilook SAR interferograms for CFAR detection of ground moving targets. IEEE Trans. Geosci. Remote Sens. **42**(4), 691–701 (2004)
8. N.R. Goodman, Statistical analysis based on a certain multivariate complex gaussian distribution (an introduction). Ann. Math. Stat. **34**(152), 152–180 (1963)
9. I.C. Sikaneta, Detection of ground moving objects with synthetic aperture radar, Ph. D. thesis. University of Ottawa, 2002
10. R. Touzi, A. Lopes, J. Bruniquel, P.W. Vachon, Coherence estimation for SAR imagery. IEEE Trans. Geosci. Remote Sens. **37**(1), 135–149 (1999)
11. R. Abdelfattah, J.M. Nicolas, Interferometric SAR coherence magnitude estimation using second kind statistics. IEEE Trans. Geosci. Remote Sens. **44**(7), 1942–1953 (2006)
12. C.J. Oliver, S. Quegan, *Understanding Synthetic Aperture Radar Images* (Artech House, Norwood, MA, 1998)
13. M. Tur, K.C. Chin, J.W. Goodman, When is speckle noise multiplicative? Appl. Opt. **21**, 1157–1159 (1982)
14. A.C. Frery, J. Jacobo-Berlles, J. Gambini, M. Mejail, Polarimetric SAR image segmentation with B-Splines and a new statistical model. Multidimension. Syst. Signal Process. **21**, 319–342 (2010)
15. A.C. Frery, A.H. Correia, C.C. Freitas, Classifying multifrequency fully polarimetric imagery with multiple sources of statistical evidence and contextual information. IEEE Trans. Geosci. Remote Sens. **45**(10), 3098–3109 (2007)
16. A.C. Frery, H.J. Muller, C.C. Freitas Yanasse, S.J. Siqueira Sant'Anna, A model for extremely heterogeneous clutter. IEEE Trans. Geosci. Remote Sens. **35**(3), 648–659 (1997)
17. J.M. Nicolas, F. Tupin, Gamma mixture modeled with second kind statistics: application to SAR image processing. Presented at the IGARSS Conference, Toronto, ON, Canada, 2002, pp. 2489–2491
18. J.M. Nicolas, Introduction to second kind statistic: Application of log-moments and log-cumulants to SAR image law analysis. Trait. Signal **19**(3), 139–167 (2002)
19. G. Moser, J. Zerubia, S.B. Serpico, SAR amplitude probability density function estimation based on a generalized Gaussian model. IEEE Trans. Image Process. **15**(6), 1429–1442 (2006)
20. C. Tison, J.M. Nicolas, F. Tupin, H. Maitre, A new statistical model for Markovian classification of urban areas in high-resolution SAR images. IEEE Trans. Geosci. Remote Sens. **42**(10), 2046–2057 (2004)
21. S.I. Gradshteyn, I.M. Ryzhik, Table of Integrals, Series, and Products, 7th edn. (Academic Press, San Diego, CA, 2007)
22. M.S. Greco, G. Gini, Statistical analysis of high-resolution SAR ground clutter data. IEEE Trans. Geosci. Remote Sens. **45**(3), 566–575 (2007)
23. Y. Delignon, W. Pieczynski, Modelling non-Rayleigh speckle distribution in SAR images. IEEE Trans. Geosci. Remote Sens. **40**(6), 1430–1435 (2002)

24. A. Achim, E.E. Kuruoglu, J. Zerubia, SAR image filtering based on the heavy-tailed Rayleigh model. IEEE Trans. Image Process. **15**(9), 2686–2693 (2006)
25. E.E. Kuruoglu, J. Zerubia, Modeling SAR images with a generalization of the Rayleigh distribution. IEEE Trans. Image Process. **13**(4), 527–533 (2004)
26. K.L.P. Vasconcellos, A.C. Frery, L.B. Silva, Improving estimation in speckled imagery. Comput. Stat. **20**(3), 503–519 (2005)
27. H. Allende, A.C. Frery, J. Galbiati, L. Pizarro, M-estimators with asymmetric influence functions: the GA0 distribution case. J. Stat. Comput. Simul. **76**(11), 941–956 (2006)
28. T.M. Cover, J.A. Thomas, *Elements of Information Theory* (Wiley, New York, 1991)

Chapter 5
Moving Vehicle Detection in Along-Track Interferometric SAR Complex Images

5.1 Introduction

In air-to-ground SAR surveillance, it is becoming increasingly desirable to develop the technique for detecting moving targets such as tanks or wheeled vehicles within strong ground clutter [1–5]. Accurate detection is crucial for the measurement of target's velocity, the estimation of location and focused-imaging [3, 5]. Therefore, SAR ground moving target detection has received a great deal of attention and been intensively investigated in the last decade [6–8].

Earlier works of the moving target detection in SAR images focused on the single-channel SAR, e.g., Fienup [4] proposed the method of segmenting a complex-valued SAR image into different patches. Additionally, the method of dividing a full-aperture into several sub-apertures was investigated [9]. Recently, with the increasing of the multi-channel and dual-channel SAR systems, some popular and well-known solutions, such as DPCA (displaced phase center antenna) [5, 10], ATI (along-track interferometry) [3, 11] and different detection metrics acquired by decomposition and transformation of the covariance matrix [5], have also been carried out. Compared with other detection methods, some experiments and applications [3, 5, 8, 11] turn out that ATI shows greater potential for detecting the slow ground moving targets.

ATI exploits the difference in the echoes of two channels observing the same scene at different times [5]. These channels are aligned in the flight direction of the SAR platform. For the stationary terrain, the two channel signals are identical and can be canceled out by computing the phase difference (i.e., the interferogram), whereas the moving targets remain in the differential data. Actually, due to the influences of "phase excursion", receiver noise and the speckle fluctuation noise, if only the interferogram's phase is utilized to detect the ground moving targets with different velocities, it is difficult to determine the detection threshold for successfully detecting all the moving targets. Fortunately, for the interesting ground moving targets, the interferogram's magnitude is helpful for the improvement of detection performance.

G. Gao, *Characterization of SAR Clutter and Its Applications to Land and Ocean Observations*, https://doi.org/10.1007/978-981-13-1020-1_5

Our objective is to present an effective and expeditious detection metric for improving the signal-to-clutter ratio (SCR) to enhance the moving targets or restrain the clutter. Meanwhile, we wish that the ground moving targets could be relatively easily detected utilizing the devised metric. Under this consideration, this section proposed a novel detection metric combing the SAR interferometric magnitude and phase, simply called IMP metric. Simultaneously, the IMP metric's statistical model, simply denoted as $\mathcal{S}^0$ distribution in this section, is derived. Additionally, the corresponding CFAR detection threshold is also presented.

5.2 The IMP Metric

5.2.1 The Characteristics of Moving Targets Compared to Stationary Clutter

In complex SAR interferogram, the magnitude and phase of the pixels corresponding to ground moving targets have the following characteristics:

(1) ***Phase comparison***: The interferogram of the moving target can be acquired by the fore-channel SAR image multiplying the complex conjugate of the aft-channel SAR image, i.e.,

$$I(t) = \mathrm{k}' \exp\left(\mathrm{j}\frac{2\pi}{\lambda} \cdot 2v_r \cdot \frac{B}{v_a}\right)\left(\sin\mathrm{c}\left(\frac{2\pi}{\lambda}\left(v_a^2((t - T_0/v) + v_r)T_s\right) \cdot T_s\right)\right)^2 \quad (5.1)$$

where k' is a constant, λ is the wavelength, B is the length of baseline, v_a is the platform velocity, v_r is the radial velocity towards the target, T_s is the aperture time, R_C is the range from the target to the flight patch, and $T_0 = -B/2v_a$. In ideal situation, due to the existence of radial velocity, the interferometric phase of the moving target is $\psi = \frac{2\pi}{\lambda} \cdot 2v_r \cdot \frac{B}{v_a}$, while that of the stationary clutter (i.e., stationary targets and natural clutter) equals to zero.

(2) ***Magnitude comparison***: The interesting ground moving targets, like the stationary targets, have stronger RCS than natural clutter. Therefore, the interferometric magnitudes corresponding to the ground moving targets and the stationary targets will be larger than natural clutter.

5.2.2 The Construction of the New Detection Metric

Based on the above analysis of the interferogram's magnitude and phase, the new detection metric can be constructed by the following steps:

Fig. 5.1 Relationship between ψ and $e^{j\psi} - 1$ in complex domain

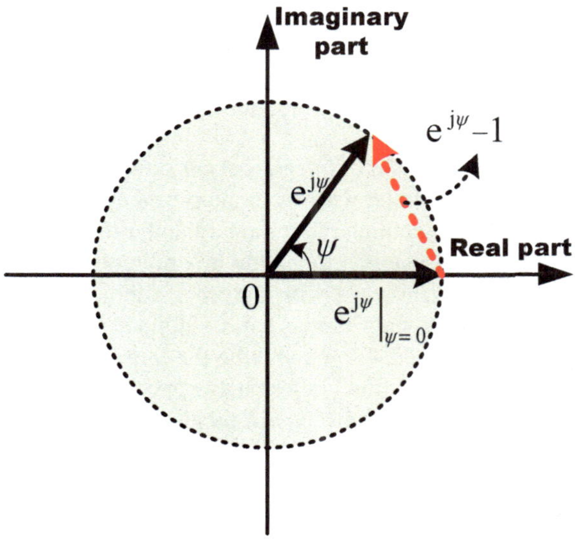

Step (1) ***The transform of interferogram's phase***: as shown in Fig. 5.1, the following transform is defined as

$$\vartheta' = e^{j\psi} - 1 \tag{5.2}$$

Obviously, ϑ' represents the difference of the interferogram's phase in the complex domain and $\psi \in (-\pi, \pi]$. The advantages that interferogram's phase ψ is transformed into variable ϑ' with (5.2) can be summarized as:

- Interferogram's phase of stationary clutter is zero, but corresponding complex interferogram is a positive real number. After transforming ψ into ϑ', the difference of interferogram's phase is zero for stationary clutter. However, the difference of interferogram's phase is not zero for moving targets, which can differentiate the moving targets and the stationary clutter effectively.
- The multilook processing can effectively suppress the influence of noise and "phase excursion", which can improve the detection performance [12]. However, considering the complexity, the multilook processing is usually replaced by the average of adjacent pixels in a window for simplification. Unfortunately, this operation will lead to the multilook interferogram phases of the moving targets are not proportional to their radial velocities because the adjacent interferometric phases of the natural clutter are also contained [12]. After transforming ψ into ϑ', moving targets is related with the difference of interferogram's phase. In another word, this transform in (5.2) can effectively eliminate the interference produced by interferogram information of stationary clutter and reflect whether moving targets exist more accurately.

Furthermore, as ϑ' is symmetrical on the real part axis, (5.2) could be simplified as:

$$\vartheta = 1 - \cos\psi \tag{5.3}$$

Step (2) ***The joint of interferogram's magnitude and phase***: Considering the natural clutter with lower interferometric magnitude compared with the targets (containing the stationary and moving targets). Therefore, the targets and the natural clutter can be differentiated by comparing the interferometric magnitude ξ with a given threshold. Furthermore, the moving targets and stationary targets can be divided in the variable ϑ domain as discussed aboved. If we combine the interferogram's magnitude ξ and phase variable ϑ, the clutter (stationary targets and natural clutter) can be rejected more effectively, and then the detection performance will be improved.

Under this consideration, the new detection metric is constructed as

$$\boldsymbol{\zeta} = \xi \cdot (1 - \cos\psi) = \xi \cdot \vartheta \tag{5.4}$$

where the variable $\boldsymbol{\zeta}$ is named IMP metric. In the variable $\boldsymbol{\zeta}$ domain, for the ground moving targets, not only their interferogram's magnitudes but their interferometric phases are larger, which results in that their corresponding values of $\boldsymbol{\zeta}$ are much more prominent; for the stationary targets, although their interferogram's magnitudes are larger, their corresponding values of the variable ϑ are lower, which leads to that their corresponding values of $\boldsymbol{\zeta}$ are relatively lower; for the natural clutter, not only their interferogram's magnitudes but their interferometric phases are lower, which induces that their corresponding values of $\boldsymbol{\zeta}$ are much smaller.

5.3 Statistical Distribution Model of IMP Metric

5.3.1 Homogeneous Area

Based on the phenomenon that the interferogram's phases corresponding to stationary scene round the zero after the multilook processing is implemented by the average of adjacent pixels in a window, $\cos\psi$ in (5.4) can be approximated with the geometric progression sum formula [13] as

$$\cos\psi \cong 1 - \frac{\psi^2}{2} \tag{5.5}$$

Consequently, $\vartheta = \psi^2/2$. the joint density of ξ and ϑ is expressed as

$$p_{\xi,\vartheta}(\xi,\vartheta)=\frac{4n^{n+1}\xi^{n}}{\pi\,\Gamma(n)\left(1-\rho^{2}\right)\sqrt{2\vartheta}}\exp\left(-\frac{2n\rho\xi}{1-\rho^{2}}(\vartheta-1)\right)K_{n-1}\left(\frac{2n\xi}{1-\rho^{2}}\right) \quad (5.6)$$

As IMP metric $\boldsymbol{\zeta}$ is the multiplication of ξ and ϑ, the PDF of $\boldsymbol{\zeta}$ can be obtained by $p_{\zeta}(\zeta)=\int_{0}^{\infty}\frac{1}{\xi}p_{\xi,\vartheta}\left(\xi,\frac{\zeta}{\xi}\right)\mathrm{d}\xi$, i.e.,

$$\begin{aligned}p_{\zeta}(\zeta)=&\frac{4n^{n+1}}{\pi\,\Gamma(n)(1-\rho^{2})\sqrt{2\zeta}}\exp\left(-\frac{2n\rho\zeta}{1-\rho^{2}}\right)\\&\cdot\int_{0}^{\infty}\xi^{n-1/2}\exp\left(\frac{2n\rho\xi}{1-\rho^{2}}\right)K_{n-1}\left(\frac{2n\xi}{1-\rho^{2}}\right)\mathrm{d}\xi\end{aligned} \quad (5.7)$$

Utilizing the following integral formula [14]:

$$\begin{aligned}\int_{0}^{\infty}x^{\mu-1}e^{-ax}K_{\nu}(bx)dx=&\frac{\sqrt{\pi}(2b)^{\nu}}{(a+b)^{\mu+\nu}}\cdot\frac{\Gamma(\mu+\nu)\Gamma(\mu-\nu)}{\Gamma(\mu+1/2)}\\&\cdot{}_{2}F_{1}\left(\mu+\nu,\nu+1/2;\mu+1/2;\frac{a-b}{a+b}\right),\ (\mathrm{Re}\,\mu>|\mathrm{Re}\,\nu|,\ \mathrm{Re}\,(a+b)>0)\end{aligned} \quad (5.8)$$

The (5.7) is given as

$$\begin{aligned}p_{\zeta}(\zeta)=&\frac{\Gamma(2n-1/2)(1+\rho)^{n-1/2}}{2\sqrt{n}\Gamma^{2}(n)(1-\rho)^{n}}\\&\cdot{}_{2}F_{1}\left(2n-1/2,n-1/2;n+1;\frac{\rho+1}{\rho-1}\right)\zeta^{-\frac{1}{2}}\exp\left(-\frac{2n\rho}{1-\rho^{2}}\zeta\right)\end{aligned} \quad (5.9)$$

where ${}_{2}F_{1}$ is the Gauss hypergeometric function. Owing to $\int_{0}^{\infty}p_{\zeta}(\zeta)\mathrm{d}\zeta=1$, the (5.9) can be simplified by the formula $\int_{0}^{\infty}\frac{e^{-qx}}{\sqrt{x}}\mathrm{d}x=\sqrt{\frac{\pi}{q}}, q>0$ [14]. Finally, the PDF of IMP metric in homogeneous area could be derived as

$$p_{\zeta}(\zeta)=\left(\frac{\upsilon_{0}}{\pi\zeta}\right)^{1/2}\exp(-\upsilon_{0}\zeta),\quad \zeta,\upsilon_{0}>0 \quad (5.10)$$

where $\upsilon_{0}=\frac{2n\rho}{1-\rho^{2}}$.

5.3.2 *The $\mathcal{S}^{0}$ Distribution*

For heterogeneous areas (such as forest and agriculture farmland areas) and extremely heterogeneous areas (such as urban areas), RCS of them have fluctuation. If IMP metric density distribution as shown in (5.10) is used to fit IMP data of heterogeneous areas or extremely heterogeneous areas, large deviations tend to generate. Herein, the multiplicative model [15] is introduced, i.e.,

$$Y_i = A_i X_i, \quad i = 1, 2 \tag{5.11}$$

where A_i represents the backscattering RCS magnitude component. $\mathcal{N}^{\mathbb{C}}(0, 1)$ denotes the complex normal distribution with expectation 0 and variance 1. X_i indicates speckle noise component and $X_i \sim \mathcal{N}^{\mathbb{C}}(0, 1)$. i is the independent receiving channel.

Furthermore, supposing that the energy of dual-channel is balanced, then the IMP metric in heterogeneous area or extremely heterogeneous areas is given by

$$\eta = A^2 \cdot \zeta = W \cdot \xi\vartheta = W \cdot \xi(1 - \cos\psi) \tag{5.12}$$

Frery et al. [16–19] recently proposed a reciprocal of the square root of a Gamma distribution for the modulating random variable A of heterogeneous clutter, i.e., $A \sim \Gamma^{-1/2}(-\alpha, \gamma)$, It can be shown that the PDF of W is

$$p_W(w) = \frac{\gamma^{-\alpha}}{\Gamma(-\alpha)} w^{\alpha-1} \exp\left(-\frac{\gamma}{w}\right), -\alpha, \gamma > 0 \tag{5.13}$$

where α ($-\alpha \in (0, \infty)$) is a shape parameter, which essentially reflects the degrees of homogeneity for processed areas. γ is a scale parameter related to the mean energy of processed areas.

Herein, within the frame of multiplicative model, combining the distribution shown in (5.10) for homogeneous area and (5.13), the PDF of the IMP metric in heterogeneous area or extremely heterogeneous areas can be given by (see Appendix 5.1).

$$p_\eta(\eta) = \frac{\sqrt{\upsilon}\Gamma(-\alpha + 1/2)}{\sqrt{\pi}\Gamma(-\alpha)} \frac{\eta^{-1/2}}{(1 + \upsilon\eta)^{-\alpha+1/2}}, \quad \eta, \upsilon, -\alpha > 0 \tag{5.14}$$

where $\upsilon = \frac{2n\rho}{\gamma(1-\rho^2)}$. The distribution characterized by the density given in Eq. (5.14) is denoted as the $\mathcal{S}^0$ distribution. It can be proved that $\mathcal{S}^0$ converges in distribution to the distribution shown in (5.10) when $-\alpha \to \infty$, $\upsilon \to 0$ and $(-\alpha) \cdot \upsilon \to 1/(2\upsilon_0)$. This property shows that either homogeneous, heterogeneous, or extremely heterogeneous statistics of the IMP metric can be treated as the outcome of the $\mathcal{S}^0$ distribution.

5.3.3 *The Parametric Estimators of the $\mathcal{S}^0$ Distribution*

The second-kind first characteristic function and the second-kind second characteristic function corresponding to the $\mathcal{S}^0$ distribution are given by (See Appendix 5.2).

$$\begin{cases} \phi_{\mathcal{S}^0}(s) = \dfrac{\Gamma(s-1/2)\Gamma(-\alpha-(s-1))}{\sqrt{\pi}\upsilon^{s-1}\Gamma(-\alpha)} \\ \varsigma_{\mathcal{S}^0}(s) = \ln\Gamma(s-1/2) + \ln\Gamma(-\alpha-(s-1)) \\ \qquad - \ln\sqrt{\pi} - (s-1)\ln\upsilon - \ln\Gamma(-\alpha) \end{cases} \tag{5.15}$$

The kth order derivative of $\varsigma_X(s)$ at $s = 1$ is the kth order second-kind cumulant named "log-cumulant". Therefore, the kth-order log-cumulant corresponding to the $\mathcal{S}^0$ distribution is described by

$$\begin{cases} \tilde{c}_1 = \Psi(1/2) - \Psi(-\alpha) - \ln\upsilon \\ \tilde{c}_k = \Psi(k-1, 1/2) + (-1)^k\Psi(k-1, -\alpha), \quad k = 2, 3, \ldots \end{cases} \tag{5.16}$$

Given a sample set $\{x_i\}$, $i \in [1, N]$, parameters υ and α in the $\mathcal{S}^0$ distribution are estimated by

$$\begin{cases} \Psi(1/2) - \Psi(-\hat{\alpha}) - \ln\hat{\upsilon} = \dfrac{1}{N}\sum\limits_{i=1}^{N}[\ln(x_i)] \\ \Psi(1, 1/2) + \Psi(1, -\hat{\alpha}) = \dfrac{1}{N}\sum\limits_{i=1}^{N}\left[(\ln(x_i) - \hat{\tilde{c}}_1)^2\right] \end{cases} \tag{5.17}$$

5.4 CFAR Detection

5.4.1 *The Threshold Derivation*

Parameters υ and α in the $\mathcal{S}^0$ distribution are estimated by (5.17). For a given value of the false alarm probability, denoted by P_{fa}, the corresponding CFAR threshold T_l for the $\mathcal{S}^0$ distribution is obtained from

$$1 - P_{fa} = \int_0^{T_l} p_\eta(\eta)\mathrm{d}\eta = \frac{\sqrt{\hat{\upsilon}}\Gamma\left(-\hat{\alpha}+1/2\right)}{\sqrt{\pi}\Gamma(-\hat{\alpha})}\int_0^{T_l}\frac{\eta^{-1/2}}{(1+\hat{\upsilon}\eta)^{-\hat{\alpha}+1/2}}\mathrm{d}\eta \tag{5.18}$$

Utilizing the formula $\int_0^u \frac{x^{\mu-1}}{(1+\beta x)^v}\mathrm{d}x = \frac{u^\mu}{\mu} {}_2F_1(v, \mu; 1+\mu; -\beta u)$ [14], the threshold T_l is given by

$$2\sqrt{\hat{\upsilon}}\Gamma(-\hat{\alpha}+1/2)T_l^{1/2}{}_2F_1\left(-\hat{\alpha}+1/2, 1/2; 3/2; -\hat{\upsilon}T_l\right) - \sqrt{\pi}\Gamma(-\hat{\alpha})(1-P_{fa}) = 0 \tag{5.19}$$

where ${}_2F_1(\cdot,\ \cdot;\ \cdot;\ \cdot)$ is the Gauss hypergeometric function, and threshold T_l can be accurately calculated by (5.19) with the help of the numerical solution.

Accordingly, for the test cell η in the image, the ground moving target is detected according to the following decision rule:

$$\eta \underset{H_0}{\overset{H_1}{\gtrless}} T_l \tag{5.20}$$

where H_1 is the hypothesis that the test cell is a ground moving target pixel, whereas H_0 is the hypothesis that the test cell is a stationary clutter pixel.

5.4.2 Detailed Flow of CFAR Detection

To summarize, as shown in Fig. 5.2, the CFAR detection method based on IMP metric for ground moving targets consists of the following steps.

Step (1) Complex conjugate multiplication and the average of adjacent pixels are implemented for registered fore and aft channel SAR images to obtain the complex multilook interferogram. First, the complex multilook interferogram is normalized. Second, ϑ image is calculated by (5.10) and magnitude ξ image is given by performing the module of the normalized complex multilook interferogram. Then, via (5.11), IMP metric image can be acquired.

Step (2) The false alarm probability is initialized. Meanwile, we estimate the parameters of the $\mathcal{S}^0$ distribution using all pixels in the corresponding IMP metric image and compute the CFAR detection threshold T_l.

Step (3) For a test cell in the IMP metric image, we obtain the binary value of the ground moving pixel by comparing the intensity of the test cell with the detection threshold T_l.

Step (4) If it is the end of IMP metric image, the final detection result is obtained. Otherwise, move on to the next pixel and repeat the process from Step (3).

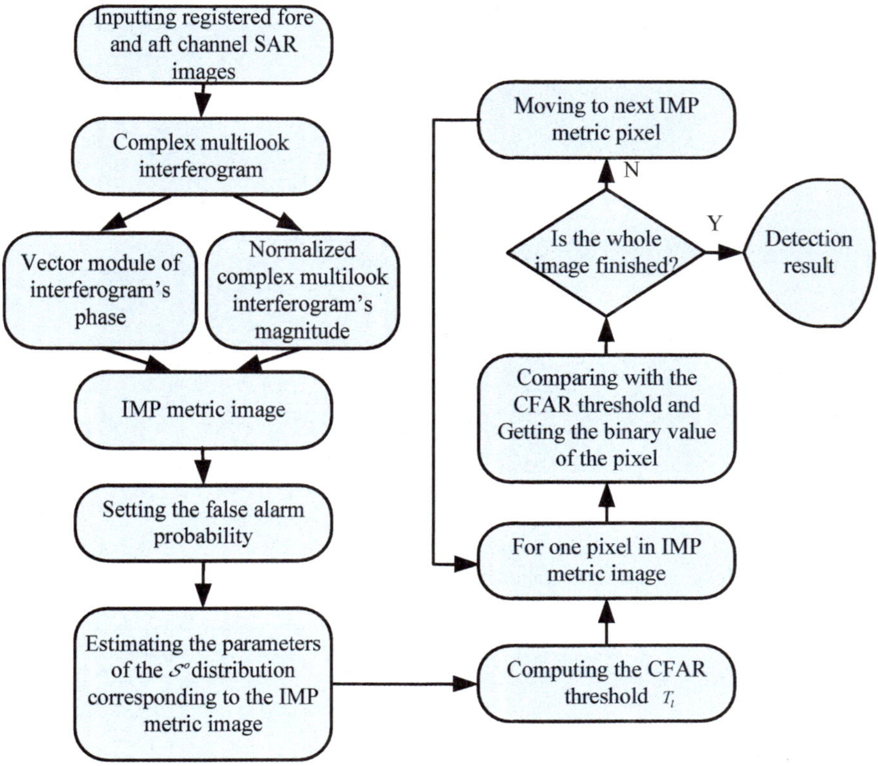

Fig. 5.2 Detailed flow of CFAR detection method based on IMP metric for ground moving targets

5.4.3 *Experimental Results*

The aim of this section is to demonstrate the capability of the presented IMP metric and the corresponding $\mathcal{S}^0$ distribution for the CFAR detection of ground moving targets based on the measured SAR data. The test dual-channel SAR data used in this study were acquired by an airborne SAR system of China in Beijing operated in X band and HH polarization, with the spatial resolution 10 m $\times$ 2 m (azimuth $\times$ range) and the size of 283 $\times$ 769 pixels. Figure 5.3 shows the fore-channel SAR magnitude image of the tested site. The horizontal and vertical axes are the directions of azimuth and range, respectively. This test site is dominated by the vegetation of shrubby and a road. Six slowly moving targets are spreading in this road.

Figure 5.4 shows the fitting results of the proposed $\mathcal{S}^0$ distribution for the IMP metric image of the area indicated in Fig. 5.3. The parameter estimations of the presented distribution are all accomplished by the proposed estimators based on the MoLC. The parameters α and υ are estimated to $\hat{\alpha} = -11.5017$ and $\hat{\upsilon} = 1.4358$, respectively. As shown in Fig. 5.4, it is clear that the almost perfect fitting result is obtained by the theoretical $\mathcal{S}^0$ distribution.

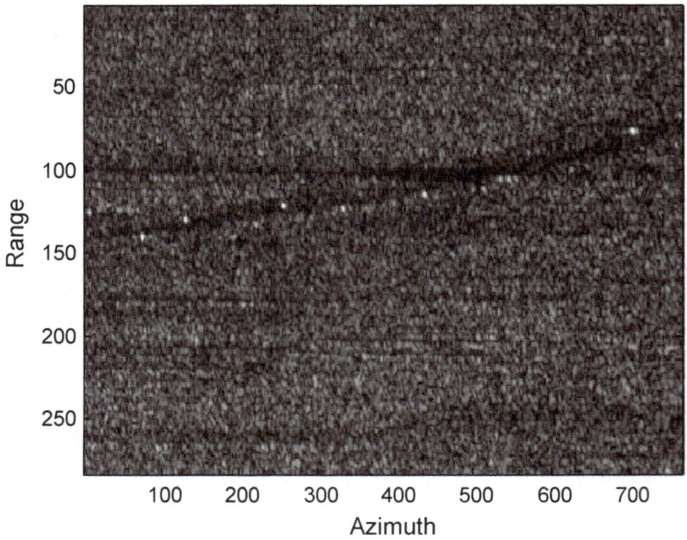

Fig. 5.3 The fore-channel SAR magnitude image of the test site

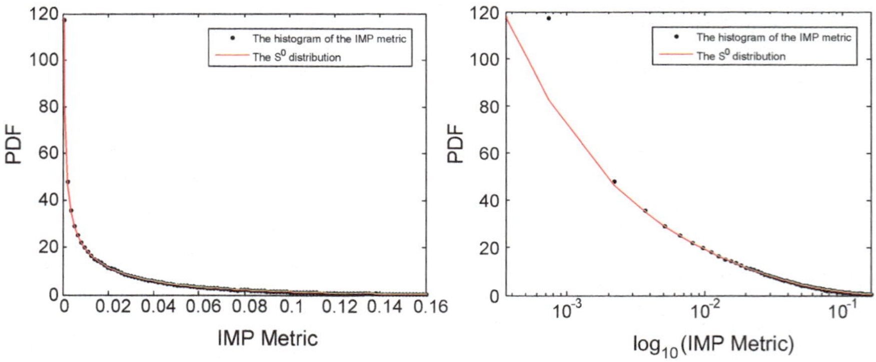

Fig. 5.4 Plots of IMP metric histogram for the test site and of the estimated $\mathcal{S}^0$ PDF

Furthermore, giving the theoretical false alarm probability $P_{fa} = 6 \times 10^{-5}$, (5.19) is used to compute the CFAR threshold based on the above-mentioned estimated values of $\hat{\alpha}$ and $\hat{\upsilon}$. Consequently, the calculated CFAR threshold is $T_l = 0.7265$. We execute the processing flow shown in Fig. 5.2 by this threshold, the detection results are shown in Fig. 5.5. It is easy to observe that all six moving targets are detected, whilst no false alarms occur, which proves the effectiveness of CFAR detection method based on IMP metric for ground moving targets.

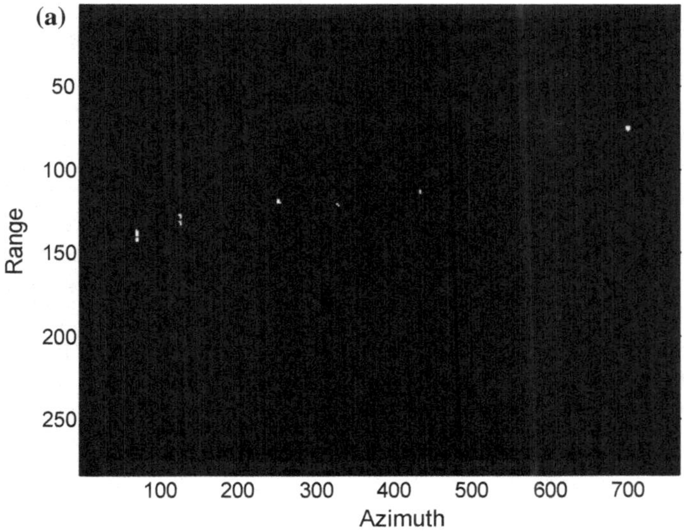

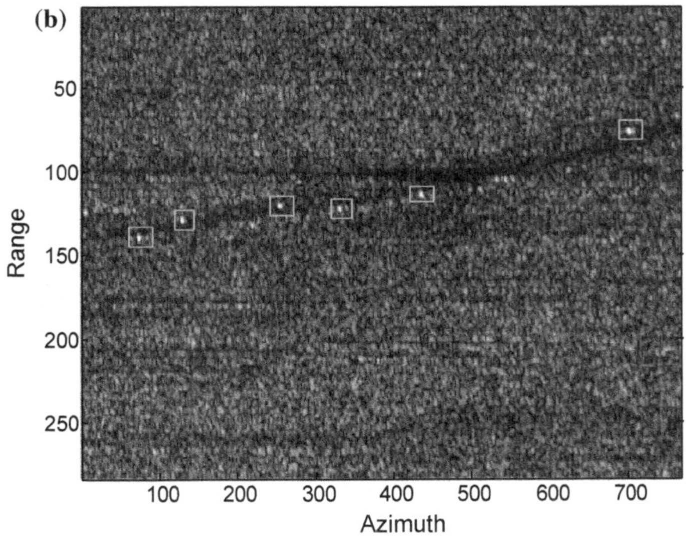

Fig. 5.5 Detection results of the proposed CFAR method with $P_{fa} = 6 \times 10^{-5}$. **a** The binary value image of detection result. **b** The detection results corresponded to original SAR image

5.5 Conclusion

Aiming at SAR ground moving target detection, this paper proposes a new joint metric of interferogram's magnitude and phase, i.e., the IMP metric, by utilizing the backscattering difference between moving targets and stationary clutter contained in the interferogram. The theoretical statistical model of the presented metric, i.e., the $\mathcal{S}^0$ distribution and the corresponding parameter estimators is also derived analytically via the multiplicative model and the reciprocal of a square root of Gamma proposed by Frery et al. [16–19]. Additionally, the CFAR detection threshold and scheme are given and designed. The experimental results of the representative site show the good performance of the proposed IMP metric, $\mathcal{S}^0$ distribution and CFAR detection method. More experiments will also be done by using more measured SAR data in our further work.

Appendix 5.1: The Derivation of the $\mathcal{S}^0$ Distribution

Via the probability density of the multiplication of two variables, the following equation can be given by combining (5.10) and (5.13)

$$p_\eta(\eta) = \int\limits_0^\infty \frac{1}{t} p_\zeta(t) p_W\left(\frac{\eta}{t}\right)\mathrm{d}t = \frac{\beta_0 \gamma^{-\alpha} \eta^{\alpha-1}}{\Gamma(-\alpha)} \int\limits_0^\infty t^{-\alpha-\frac{1}{2}} \exp\left[-\left(\frac{2n\rho}{1-\rho^2} + \frac{\gamma}{\eta}\right)t\right]\mathrm{d}t \tag{5.21}$$

where $\beta_0 = \sqrt{\frac{2n\rho}{\pi(1-\rho^2)}}$. Utilizing the formula shown as

$$\int\limits_0^\infty x^{m-1} \mathrm{e}^{-(a+1)x} \mathrm{d}x = \frac{\Gamma(m)}{(a+1)^m}, m > 0, a > -1 \tag{5.22}$$

Replacing the variables by $m = -\alpha + 1/2$, $a + 1 = \frac{2n\rho}{1-\rho^2} + \frac{\gamma}{\eta}$, (5.22) is completely identical to the PDF of $\mathcal{S}^0$ shown in (5.14).

Appendix 5.2: The Second-Kind First Characteristic Function of the $\mathcal{S}^0$ Distribution

Taking (5.14) into the expression of Mellin transform [20–23], we obtain

$$\begin{aligned}\phi_{S^0}(s) &= \int_0^\infty \eta^{s-1} p_\eta(\eta)\mathrm{d}\eta \\ &= \frac{\sqrt{\upsilon}\,\Gamma(-\alpha+1/2)}{\sqrt{\pi}\,\Gamma(-\alpha)}\int_0^\infty \frac{\eta^{s-3/2}}{(1+\upsilon\eta)^{-\alpha+1/2}}\mathrm{d}\eta\end{aligned} \tag{5.23}$$

Via the following integral formula

$$\int_0^\infty \frac{t^{\mu-1}}{(1+bt^a)^{\mu+l}}\mathrm{d}t = a^{-1}b^{-\frac{\mu}{a}}B\left(\frac{\mu}{a},\mu+l-\frac{\mu}{a}\right), a, b > 0 \tag{5.24}$$

Let $a = 1, b = \upsilon, \mu = s - 1/2$, and $l = -\alpha - (s-1)$, (5.23) becomes as

$$\phi_{S^0}(s) = \frac{\Gamma(s-1/2)\Gamma(-\alpha-(s-1))}{\sqrt{\pi}\upsilon^{s-1}\Gamma(-\alpha)} \tag{5.25}$$

References

1. D.C. Maori, J. Klare, A.R. Brenner, J.H.G. Ender, Wide-area traffic monitoring with the SAR/GMTI system PAMIR. IEEE Trans. Geosci. Remote Sens. **46**(10), 3019–3030 (2008)
2. T. Wang, Z. Bao, Z. Zhang, J. Ding, Improving coherence of complex image pairs obtained by along-track bistatic SARs using range-azimuth prefiltering. IEEE Trans. Geosci. Remote Sens. **46**(1), 3–13 (2008)
3. E. Chapin, C.W. Chen, Along-track interferometry for ground moving target indication. IEEE Aerosp. Electron. Syst. Mag. **23**(6), 19–24 (2008)
4. J.R. Fienup, Detecting moving targets in SAR imagery by focusing. IEEE Trans. Aerosp. Electron. Syst. **37**(3), 794–809 (2001)
5. I.C. Sikaneta, Detection of ground moving objects with synthetic aperture radar, Ph. D. thesis. University of Ottawa, 2002
6. J.N. Entzminger, JointSTARS and GMTI: past, present and future. IEEE Trans. Aerosp. Electron. Syst. **35**(2), 748–761 (1999)
7. H. Steyskal, J.K. Schindler, P. Franchi, R.J. Mailloux, Pattern synthesis for TechSat21-A distributed space-based radar system. IEEE Antennas Propag. Mag. **45**(4), 19–25 (2003)
8. C. Livingstone, A. Thompson, The moving object detection experiment on RADARSAT-2. Can. J. Remote. Sens. **30**(3), 355–368 (2004)
9. A. Moreira, Real-time synthetic aperture radar (SAR) processing with a new sub-aperture approach. IEEE Trans. Geosci. Remote Sens. **30**(4), 714–722 (1992)
10. C.H. Gierull, Statistics of SAR interferograms with application to moving target detection, *DREO Technical Report TR 2001-045*, Defense Research Establishment Ottawa, Department of National Defense, Ottawa, Canada, 2001

11. C.H. Gierull, Statistical analysis of multilook SAR interferograms for CFAR detection of ground moving targets. IEEE Trans. Geosci. Remote Sens. **42**(4), 691–701 (2004)
12. J.S. Lee, K.W. Hoppel, S.A. Mango, A.R. Miller, Intensity and phase statistics of multilook polarimetric and interferometric SAR imagery. IEEE Trans. Geosci. Remote Sens. **32**(5), 1017–1028 (1994)
13. E. Previato, *Dictionary of Applied Math for Engineers and Scientists* (CRC Press, London, 2003)
14. S.I. Gradshteyn, I.M. Ryzhik, *Table of Integrals, Series, and Products*. 7th edn. (Academic Press, San Diego, CA, 2007)
15. C.J. Oliver, S. Quegan, *Understanding Synthetic Aperture Radar Images* (Artech House, Norwood, MA, 1998)
16. A.C. Frery, H.J. Muller, C.C. Freitas Yanasse, S.J. Siqueira Sant'Anna, A model for extremely heterogeneous clutter. IEEE Trans. Geosci. Remote Sens. **35**(3), 648–659 (1997)
17. A.C. Frery, J. Jacobo-Berlles, J. Gambini, M. Mejail, Polarimetric SAR image segmentation with B-Splines and a new statistical model. Multidimension. Syst. Signal Process. **21**, 319–342 (2010)
18. A.C. Frery, A.H. Correia, C.C. Freitas, Classifying multifrequency fully polarimetric imagery with multiple sources of statistical evidence and contextual information. IEEE Trans. Geosci. Remote Sens. **45**(10), 3098–3109 (2007)
19. C.C. Freitas, A.C. Frery, A.H. Correia, The polarimetric G distribution for SAR data analysis. Environmetrics **16**(1), 13–31 (2005)
20. N.R. Goodman, Statistical analysis based on a certain multivariate complex gaussian distribution (an introduction). Ann. Math. Stat. **34**(152), 152–180 (1963)
21. J.M. Nicolas, F. Tupin, Gamma mixture modeled with second kind statistics: Application to SAR image processing. Paper presented at the IGARSS Conference, Toronto, ON, Canada, 2002, pp. 2489–2491
22. J.M. Nicolas, Introduction to second kind statistic: Application of log-moments and log-cumulants to SAR image law analysis. Trait. Signal **19**(3), 139–167 (2002)
23. R. Abdelfattah, J.M. Nicolas, Interferometric SAR coherence magnitude estimation using second kind statistics. IEEE Trans. Geosci. Remote Sens. **44**(7), 1942–1953 (2006)

Chapter 6
Statistical Modeling and Target Detection of PolSAR Images

6.1 Introduction

The interpretation of the polarimetric synthetic radar (PolSAR) image, which can be fully represented by the polarimetric covariance matrix, is of considerable current interest. To interpret the PolSAR image effectively, many researches on image processing like denoising, segmentation, classification and edge detection algorithms [1–5] are carried out. Of great importance is the accurate description of statistical properties of PolSAR image with which almost all the above techniques associate. For this purpose, several distributions have been proposed with the development of SAR systems.

The complex Wishart distribution proposed by Goodman [6] was introduced for analyzing the PolSAR data in the early studies. However, with the deepening of study, many researches [7–12] have shown that this model is only agree with the homogeneous regions well and less viable when considering the heterogeneous areas. To solve this problem, some other studies were carried out and the remarkable achievement was the utilization of the multiplicative model, also called product model, from which several well-known models, $\mathcal{K}_P$ [7, 8, 10], $\mathcal{G}_p$ [11] and KummerU distribution [12], had been derived.

In the multiplicative model, the observed multilook covariance matrix is assumed to be the product of two independent random factors, namely the real and positive backscatter/texture and the Wishart distributed speckle [11]. Under the framework, the $\mathcal{K}_P$, $\mathcal{G}_p$ and KummerU distribution can be derived with the gamma, generalized inverse Gaussian and Fisher distribution modeling the backscatter, respectively [7–12]. $\mathcal{G}_p$ distribution, the extension of the famous univariate $\mathcal{G}$ distribution [13] and has the aforementioned $\mathcal{K}_P$ distribution as special case, is not practical for lack of efficient estimator. However, another special case of $\mathcal{G}_P$, the $\mathcal{G}_P^0$ distribution [11], with the backscatter modeled by inverse gamma distribution, does not have this difficult and shows better performance than $\mathcal{K}_P$. The KummerU distribution proposed especially for the extremely heterogeneous areas has $\mathcal{K}_P$ and $\mathcal{G}_P^0$ distributions as special cases, and has been successfully applied to the segmentation algorithms

G. Gao, *Characterization of SAR Clutter and Its Applications to Land and Ocean Observations*, https://doi.org/10.1007/978-981-13-1020-1_6

[12, 14]. However, its parameter estimator seems to be of high mathematical complexity. Besides, it is found that parts of observed PolSAR data is outside the KummerU distribution domain defined in the famous $\tilde{\kappa}_3 \sim \tilde{\kappa}_2$ diagram, which means the distribution is not fit for the "outside" data, and it may be severe for the complex region in the high resolution images.

Generalized gamma distribution (GΓD) [15] is a flexible empirical model which forms a large variety of alternative distributions including the gamma and the inverse gamma distribution, and cover a larger range than the Fisher distribution in the $\tilde{\kappa}_3 \sim \tilde{\kappa}_2$ diagram. Considering the advantages of this model, in this chapter, we try to employ the (GΓD) to model the backscatter in the framework of multiplicative model and derive a novel distribution for the multilook polarimetric covariance matrix.

Recently, the method of log-cumulants (MoLC) proposed by Nicolas shows its outstanding advantage in parameter estimating for distributions defined on the positive interval $\mathbb{R}^+$ [15–21]. In addition, an extension version of this method in the cone of complex positive definite matrices, named method of matrix log-cumulants (MoMLC), has been propose by Anfinsen et al. and shown to be well adapt to analyze the multilook polarimetric SAR data [22, 23]. In this chapter, based on the MoMLC, we derive the parameter estimator of the proposed model.

6.2 Multiplicative Model for Covariance Matrix

6.2.1 Multilook PolSAR Data

A full-polarimetric SAR measures the complex scattering matrix **S** containing the polarimetric information of the scenes with the form as follows:

$$\mathbf{S} = \begin{bmatrix} S_{HH} & S_{HV} \\ S_{VH} & S_{VV} \end{bmatrix} \tag{6.1}$$

where "H" and "V" represent the horizontal and vertical polarization respectively, and $S_{ij}(i, j = H, V)$ is the scattering coefficient for transmitting in j polarization and receiving in i polarization. For the monostatic SAR which usually employs the same antenna for both transmitting and receiving, it can be considered that the cross-polarization components S_{HV} and S_{VH} are equal [11], so the scattering matrix **S** can be simplified with the vector as follows:

$$\mathbf{Z} = \begin{bmatrix} S_{HH} \\ S_{HV} \\ S_{VV} \end{bmatrix} \tag{6.2}$$

For the purpose of suppressing the noise or compressing the image, the multilook processing is often carried out by averaging the sample scattering covariance matrix. The n-looks covariance matrix $\mathbf{Z}^{(n)}$ is formed as follows [9–11]:

$$\mathbf{Z}^{(n)} = \frac{1}{n}\sum_{k=1}^{n}\mathbf{Z}(k)\mathbf{Z}(k)^{H} \tag{6.3}$$

where the superscript "H" denotes complex conjugate transpose operator.

6.2.2 Multiplicative Model

The multiplicative mode assumes that the single look scattering vector $\mathbf{Z}$ is the result of the product of two independent factors, namely, $\mathbf{Z} = X^{1/2}\mathbf{Y}$ with X a unit mean random variable representing the backscatter and $\mathbf{Y}$ distributing the speckle noise. Usually, the number of looks n is small. In this case, we can assume that the backscatter component of $\mathbf{Z}(k)$, noted by $X(k)$, is independent of k [10]. So the n-looks covariance matrix can be rewritten as:

$$\mathbf{Z}^{(n)} = \frac{X}{n}\sum_{k=1}^{n}\mathbf{Y}(k)\mathbf{Y}(k)^{H} = X\mathbf{Y}^{(n)} \tag{6.4}$$

where $\mathbf{Y}^{(n)}$ is the n-looks covariance matrix of the speckle. According to the theory of probability, the PDF of the covariance matrix $f_{\mathbf{Z}^{(n)}}(\mathbf{z})$ can be calculated by:

$$\begin{aligned} f_{\mathbf{Z}^{(n)}}(\mathbf{z}) &= \int_{0}^{\infty} f_{\mathbf{Z}^{(n)}}(\mathbf{z}|x) f_X(x)dx \\ &= \int_{0}^{\infty} x^{-d^2} f_{\mathbf{Y}^{(n)}}\left(\frac{\mathbf{z}}{x}\right) f_X(x)dx \end{aligned} \tag{6.5}$$

where d is the dimension of the scattering vector $\mathbf{Y}$ or $\mathbf{Z}$, and $f_X(x)$ is the density of the backscatter X. Under the multiplicative model, the speckle component assumed to be fully developed follows a multivariate complex distribution [6, 10],

$$f_{\mathbf{Y}}(\mathbf{y}) = \frac{1}{\pi^{d}|\mathbf{C}|}\exp\left(-\mathbf{y}^{H}\mathbf{C}^{-1}\mathbf{y}\right) \tag{6.6}$$

where $\mathbf{C} = E(\mathbf{Y}\mathbf{Y}^{H})$, "$E$" means the expected value and "$|\cdot|$" denote the determinant operator. In this case, it can be derived that $\mathbf{Y}^{(n)}$ follows the Wishart distribution [6], noted $\mathbf{Y}^{(n)} \sim \mathcal{W}(\mathbf{C}, n)$, with its PDF given by

$$f_{\mathbf{Y}^{(n)}}(\mathbf{y}) = \frac{n^{nd}|\mathbf{y}|^{n-d}\exp[-nTr(\mathbf{C}^{-1}\mathbf{y})]}{\Gamma_d(n)|\mathbf{C}|^n} \tag{6.7}$$

where $\mathbf{C} = E(\mathbf{YY}^H) = E(\mathbf{ZZ}^H)$, "$Tr(\cdot)$" denotes the trace operator, $\Gamma_d(s) = \pi^{\frac{d(d-1)}{2}} \prod_{i=0}^{d-1} \Gamma(s-i)$ called multivariate gamma function in [23], and $\Gamma(\cdot)$ denotes the gamma function.

It can be seen that different distributions chosen for backscatter X would lead to different models of the $\mathbf{Z}^{(n)}$. X is assumed to be a constant that $f_X(x) = 1$ when the areas is homogeneous, which leads to $\mathbf{Z}^{(n)}$ hold the law of Wishart distribution as $\mathbf{Y}^{(n)}$ [11]. For modeling the backscatter in heterogeneous areas, several distributions as mentioned above have been chosen, such as gamma, inverse gamma and Fisher distribution that give rise to the well-known distributions as $\mathcal{K}_P$, $\mathcal{G}_P^0$ and KummerU respectively.

6.3 Statistical Modeling of PolSAR Images with Generalized Gamma Distribution for Backscatter

6.3.1 *Advantage of GΓD*

The $\tilde{\kappa}_3 \sim \tilde{\kappa}_2$ diagram, the representation of the third order log-cumulant $\tilde{\kappa}_3$ (its definition would be given later) versus the second order log-cumulant $\tilde{\kappa}_2$, shows its advantage to describe the distributions defined on $\mathbb{R}^+$ compared to the traditional Person diagram [17], from which we can assess the flexibility of certain model and select a relatively appropriate distribution for the given data [15, 17, 18]. Figure 6.1 illustrates the log-normal, gamma, inverse gamma, univariate $\mathcal{K}$ [13], Fisher distributions [18] as well as the GΓD in the $\tilde{\kappa}_3 \sim \tilde{\kappa}_2$ diagram, of which the mathematic expressions of $\tilde{\kappa}_2$ and $\tilde{\kappa}_3$ are presented in [17–19]. Generally speaking, a wider range the model covers implies a better model capability. Visually can we see from Fig. 6.1 that the GΓD covers a wider subspace than the other considering distributions.

In order to show the modeling capacity of GΓD for actual SAR data, an example is taken on a HH-polarized 4 looks L-band NASA/JPL AIRSAR intensity image (Fig. 6.2a). The empirical $\tilde{\kappa}_2$ and $\tilde{\kappa}_3$ were calculated by a 15×15 window over the image and the empirical $(\tilde{\kappa}_3, \tilde{\kappa}_2)$ points (black points) as well as the fields of Fisher and GΓD were then plotted (Fig. 6.2b). It can be observed that most of the empirical $(\tilde{\kappa}_3, \tilde{\kappa}_2)$ points belong to the GΓD field, more than that to the Fisher range. Three regions corresponding to the exceptional points are indicated by the rectangles in Fig. 6.2a. Similar experiments were carried out in [14, 15]. These experimental results suggest the GΓD owns a higher flexibility than the Fisher, gamma, inverse gamma distribution, etc. So, it would be attractive to employ the GΓD to describe the backscatter in the multiplicative model for PolSAR data.

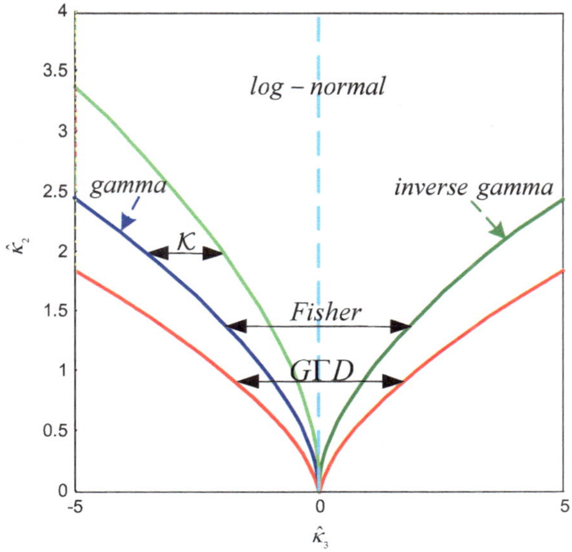

Fig. 6.1 Distributions displayed in the $\tilde{\kappa}_3 \sim \tilde{\kappa}_2$ diagram

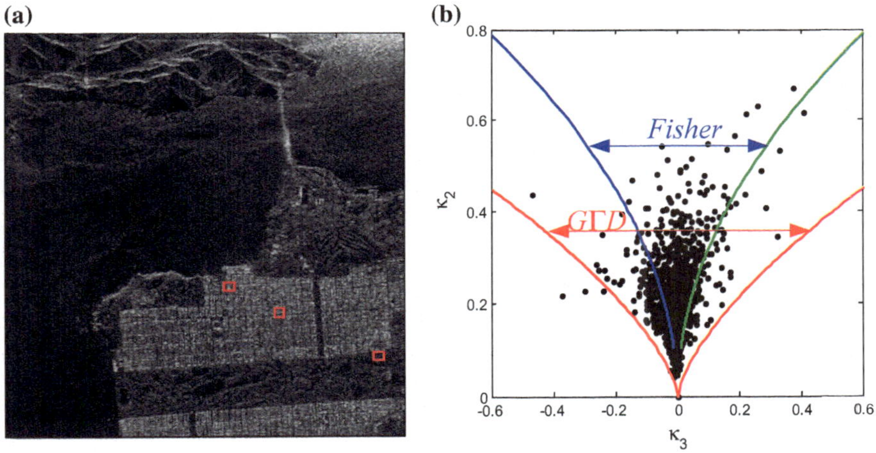

Fig. 6.2 **a** The L-Band HH polarized "San Francisco" intensity image and **b** corresponding empirical $(\tilde{\kappa}_3, \tilde{\kappa}_2)$ calculated with a 15 × 15 window

6.3.2 The Compound Model

As the distribution of the multiplicative model is the result of density of speckle compounds with that of backscatter, it is called as compound distribution [22, 23]. By substituting (6.7) and (3.1) into (6.5), the PDF of the compound model can be derived as

$$f_{\mathbf{Z}^{(n)}}(\mathbf{z}) = \frac{|\nu| n^{nd} |\mathbf{z}|^{n-d} \xi^{\kappa\nu}}{\Gamma_d(n)|\mathbf{C}|^n \Gamma(\kappa)} \cdot \int_0^{\infty} x^{\kappa\nu - nd - 1} \exp\left\{ -\frac{nTr(\mathbf{C}^{-1}\mathbf{z})}{x} - (\xi x)^{\nu} \right\} dx \tag{6.8}$$

As the GΓD is used to model the backscatter, for brief expressing, we note this compound model as polarimetric compound generalized gamma ($CG\Gamma_P$) distribution in this chapter. By setting $\nu = 1$ and $\nu = -1$, two special cases of $CG\Gamma_P$ law can be derived as (6.9) and (6.10) respectively (see Appendix 6.1).

$$f_{\mathbf{Z}^{(n)}}(\mathbf{z}) = 2\frac{|\mathbf{z}|^{n-d}(n\kappa)^{(\kappa+nd)/2}}{\Gamma_d(n)|\mathbf{C}|^n \Gamma(\kappa)} \cdot \frac{K_{\kappa-nd}\left(2\sqrt{\kappa n Tr(\mathbf{C}^{-1}\mathbf{z})}\right)}{\left(Tr(\mathbf{C}^{-1}\mathbf{z})\right)^{-(\kappa-nd)/2}} \tag{6.9}$$

$$f_{\mathbf{Z}^{(n)}}(\mathbf{z}) = \frac{n^{nd}|\mathbf{z}|^{n-d}(\kappa-1)^{\kappa}\Gamma(\kappa+nd)}{\Gamma_d(n)|\mathbf{C}|^n\Gamma(\kappa)} \cdot \left(nTr(\mathbf{C}^{-1}\mathbf{z}) + \kappa - 1\right)^{-\kappa-nd} \tag{6.10}$$

where $K_m(\cdot)$ is the mth-order modified Bessel function of the second kind. The former is $\mathcal{K}_P$ distribution presented in [10] and latter has the same expression of $\mathcal{G}_P^0$ in [11] by making $\kappa = -\alpha$. It has been proved that both (6.9) and (6.10) would yield the Wishart distribution when $\kappa \to \infty$ [11, 24]. The relationship between $CG\Gamma_P$ model and its forementioned special cases is represented in (6.11), where "$\xrightarrow{D}$" means the convergence in distribution. It implies the potential of the $CG\Gamma_P$ law in modeling the PolSAR data.

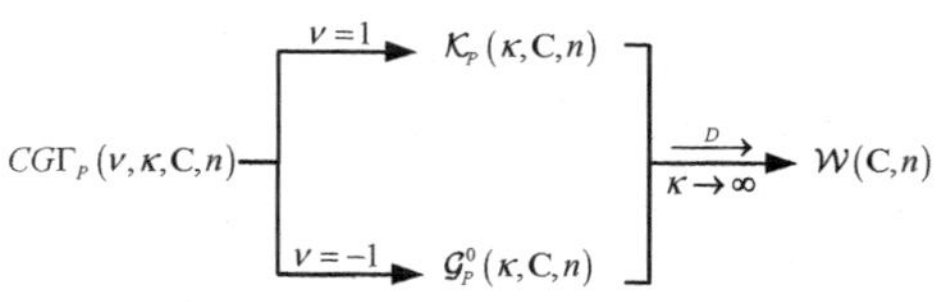

(6.11)

The intensity distribution of each channel j ($j = HH, HV, VV$) can be obtained from (6.8) by letting $d = 1$.

$$f_{I_j}(z) = \left(\frac{n}{\lambda_j}\right)^n \frac{|\nu|\xi^{\kappa\nu}z^{n-1}}{\Gamma(n)\Gamma(\kappa)} \int_0^{\infty} x^{\kappa\nu-n-1} \exp\left\{-\frac{nz}{x\lambda_j} - (\xi x)^{\nu}\right\} dx \tag{6.12}$$

where $\lambda_j = E[z]$ denoting the mean value of channel j. Similarly, (6.12) would lead to the univariate $\mathcal{K}$ and $\mathcal{G}^0$ distribution when $\kappa \to \infty$ with $\nu = 1$ and $\nu = -1$

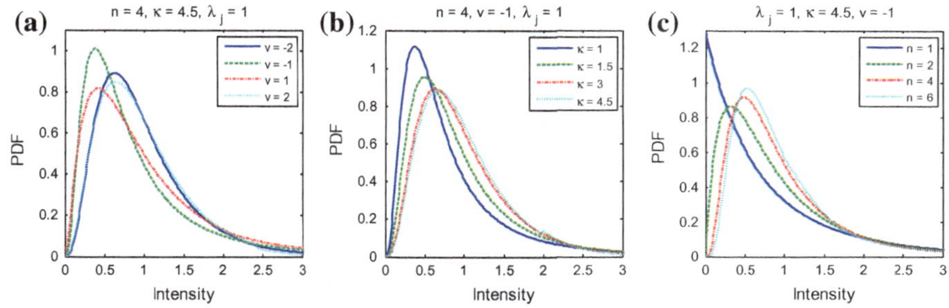

Fig. 6.3 Marginal PDF of $CG\Gamma_P$ with $\lambda_j = 1$ and different values of v, κ and n: **a** $n = 4$, $\kappa = 4.5$, $v = \{-2, -1, 1, 2\}$, **b** $n = 4$, $v = -1$, $\kappa = \{1, 1.5, 3, 4.5\}$, **c** $\kappa = 4.5$, $v = -1$, $n = \{1, 2, 4, 6\}$

respectively. Figure 6.3 shows the plots of $f_{I_j}(z)$ with the mean value $\lambda_j = 1$ and different values of v, κ and n.

6.3.3 *Estimator Based on Method of Matrix Log-Cumulants*

The parameters of $CG\Gamma_P$, namely, $\mathbf{C}$, n, κ, v, need to be estimated from the empirical PloSAR data. The matrix $\mathbf{C}$ can be estimated by averaging the sample covariance matrices $\mathbf{Z}_i^{(n)}$, $(i = 1, 2, \ldots)$ [11]. Referring to the other parameters, the application of the standard estimations, i.e., method of moments (MoM) and maximum like (ML) methodology, turns out to be a hard task for it would lead to a high nonlinear problem.

Mellin transform is an useful tool for forming the parameter estimator [15–21]. In view of the density for covariance matrix, the matrix-variate Mellin transform has been introduced and the Method of matrix Log-cumulants (MoMLC) is proposed [22, 23]. In this chapter, we apply the MoMLC in estimating the parameters of $CG\Gamma_P$.

6.3.3.1 The Mellin Kind Cumulants and It's Matrix Versions

Assuming u is a positive random variate with its PDF as $f_u(u)$, the first Mellin kind characteristic function $\phi_u(s)$ of u is defined as [17, 20]

$$\phi_u(s) = \mathcal{M}\{f_u(u)\}(s) = \int_0^\infty u^{s-1} f_u(u) du, \quad s \in \mathbb{C} \tag{6.13}$$

where $\mathcal{M}$ is the univariate Mellin transform defined on the positive real interval, $\mathbb{C}$ denotes the complex plane. The second Mellin kind characteristic function, also named as Mellin kind cumulant generating function in [23], $\varphi_u(s)$ is defined as the logarithm transformation of $\phi_u(s)$, i.e.,

$$\varphi_u(s) = \ln \phi_u(s) \tag{6.14}$$

The mth-order Mellin kind cumulants (log-cumulants) is given by

$$\tilde{\kappa}_m\{u\} = \frac{d^m}{ds^m}\varphi_u(s)\bigg|_{s=1}, \quad m = 1, 2, \ldots \tag{6.15}$$

Let $\mathbf{U}$ be a positive definite d $\times$ d Hermitian matrix with its PDF as $f_{\mathbf{U}}(\mathbf{U})$, the Mellin kind characteristic functions of $\mathbf{U}$, denoted as $\phi_{\mathbf{U}}(s)$, is defined as [22, 23, 25].

$$\phi_{\mathbf{U}}(s) = \mathcal{M}\{f_{\mathbf{U}}(\mathbf{U})\}(s) = \int\limits_{\Omega_+} |\mathbf{U}|^{s-d} f_{\mathbf{U}}(\mathbf{U}) d\mathbf{U}, s \in \mathbb{C} \tag{6.16}$$

where $\mathcal{M}$ is the matrix-variate Mellin transform.

Analogous to the univariate cases, the Mellin kind cumulant generating function $\varphi_{\mathbf{U}}(s)$ and m-order matrix log-cumulants (MLC) $\kappa_m\{\mathbf{U}\}$ of $\mathbf{U}$ are given respectively by [23].

$$\varphi_{\mathbf{U}}(s) = \ln \phi_{\mathbf{U}}(s) \tag{6.17}$$

$$\tilde{\kappa}_m\{\mathbf{U}\} = \frac{d^m}{ds^m}\varphi_{\mathbf{U}}(s)\bigg|_{s=d}, \quad m = 1, 2, \ldots \tag{6.18}$$

A definition and a theorem with its corollary is needed for form the estimator for $CG\Gamma_P$.

Definition (Matrix-Variate Mellin Convolution): Assuming $g(\mathbf{U})$ and $h(\mathbf{U})$, two functions defined on the cone of positive definite complex Hermitian matrices, are symmetric, take $g(\mathbf{U})$ for an example, i.e.,

$$g(\mathbf{UV}) = g(\mathbf{VU}) = g(\mathbf{U}^{1/2}\mathbf{V}\mathbf{U}^{1/2}) = g(\mathbf{V}^{1/2}\mathbf{U}\mathbf{V}^{1/2}), \tag{6.19}$$

The matrix-variate Mellin convolution of $g(\mathbf{U})$ and $h(\mathbf{U})$ is defined as [23]

$$\begin{aligned}(g \hat{\star} h)(\mathbf{U}) &= \int\limits_{\Omega_+} |\mathbf{V}|^{-d} g\left(\mathbf{V}^{-1/2}\mathbf{U}\mathbf{V}^{-1/2}\right) h(\mathbf{V}) d\mathbf{V} \\ &= \int\limits_{\Omega_+} |\mathbf{V}|^{-d} h\left(\mathbf{V}^{-1/2}\mathbf{U}\mathbf{V}^{-1/2}\right) g(\mathbf{V}) d\mathbf{V}\end{aligned} \tag{6.20}$$

Theorem For the functions $g(\mathbf{U})$ and $h(\mathbf{U})$ defined above, we have [23]

$$\mathcal{M}\{(g\hat{\star}h)(\mathbf{U})\}(s) = \mathcal{M}\{g(\mathbf{U})\}(s) \cdot \mathcal{M}\{h(\mathbf{U})\}(s). \tag{6.21}$$

Corollary Assuming the positive definite Hermitian matrices **U**, **V** and **W** satisfy that

$$\mathcal{M}\{(g\hat{\star}h)(\mathbf{W})\}(s) = \mathcal{M}\{g(\mathbf{U})\}(s) \cdot \mathcal{M}\{h(\mathbf{V})\}(s) \tag{6.22}$$

and all of their MLCs exist, it can be derived that [23]

$$\tilde{\kappa}_m\{\mathbf{W}\} = \tilde{\kappa}_m\{\mathbf{U}\} + \tilde{\kappa}_m\{\mathbf{V}\}, \quad m = 1, 2, \ldots. \tag{6.23}$$

6.3.3.2 Parameter Estimator for $CG\Gamma_p$ Base on MoMLC

By virtue of Eq. (6.4), the matrix $\mathbf{Z}^{(n)}$ can be rewritten as the product of two positive definite Hermitian matrices $\mathbf{X}$ and $\mathbf{Y}^{(n)}$, namely, $\mathbf{Z}^{(n)} = \mathbf{X}\mathbf{Y}^{(n)}$, where $\mathbf{X} = X\mathbf{E}_d$ and $\mathbf{E}_d$ is the d-dimension identity matrix. It can be proved that the PDFs of $\mathbf{Z}^{(n)}$, $\mathbf{X}$ and $\mathbf{Y}^{(n)}$, denoted by $f_{\mathbf{Z}^{(n)}}(\mathbf{z}), f_{\mathbf{X}}(\mathbf{x})$ and $f_{\mathbf{Y}^{(n)}}(\mathbf{y})$ respectively, hold the law that [23]

$$\begin{aligned}\mathcal{M}\{f_{\mathbf{Z}^{(n)}}(\mathbf{z})\}(s) =& \mathcal{M}\{(f_{\mathbf{X}}\hat{\star}f_{\mathbf{Y}^{(n)}})(\mathbf{z})\}(s)\\ =& \mathcal{M}\{f_{\mathbf{X}}(\mathbf{x})\}(s) \cdot \mathcal{M}\{f_{\mathbf{Y}^{(n)}}(\mathbf{y})\}(s).\end{aligned} \tag{6.24}$$

It follows from the *Corollary* that

$$\tilde{\kappa}_m\{\mathbf{z}\} = \tilde{\kappa}_m\{\mathbf{x}\} + \tilde{\kappa}_m\{\mathbf{y}\}, \quad m = 1, 2, \ldots \tag{6.25}$$

So the problem of calculating the MLCs of $\mathbf{Z}^{(n)}$ turns to be the issue of computing the MLCs of $\mathbf{X}$ and $\mathbf{Y}^{(n)}$. By virtue of the relevant definitions above, the MLCs of $\mathbf{Y}^{(n)}$ can be calculated as

$$\tilde{\kappa}_m\{\mathbf{y}\} = \begin{cases} \Psi_d(0, n) + \ln|\mathbf{C}| - d\ln(n) & m = 1 \\ \Psi_d(m-1, n) & m > 1 \end{cases} \tag{6.26}$$

with the m-order multivariate polygamma function $\Psi_d(m, t)$ defined as [22, 23]

$$\Psi_d(m, t) = \frac{d^{m+1}}{dz^{m+1}} \ln(\Gamma_d(t)) = \sum_{i=0}^{d-1} \Psi(m, t-i), \quad m = 1, 2, \ldots \tag{6.27}$$

where $\Psi(m, t) = \frac{d^m}{dn^m} \ln \Gamma(t)$ is the m-order polygamma function.

Substituting (3.1) to (6.13), the second kind character function of X follows the unitary mean GΓD can be derived as

$$\phi_X(s) = \frac{\xi^{1-s}}{\Gamma(\kappa)}\Gamma\left(\kappa + \frac{s-1}{\nu}\right) \tag{6.28}$$

According to (6.14) and (6.15), we have

$$\tilde{\kappa}_m(X) = \begin{cases} \Psi(\kappa)/\nu + \ln\Gamma(\kappa) - \ln\Gamma(\kappa + 1/\nu) & m = 1 \\ \Psi(m-1,\kappa)/\nu^m & m > 1 \end{cases} \tag{6.29}$$

In [23], it has been proved that MLCs of $\mathbf{X} = X\mathbf{E}_d$ can be calculated by $\tilde{\kappa}_m(\mathbf{X}) = d^m\tilde{\kappa}_m(X)$. So the MLCs of $\mathbf{Z}^{(n)}$ are given by

$$\tilde{\kappa}_m\{\mathbf{z}\} = \begin{cases} \Psi_d(n) + \ln|\mathbf{C}| + d\left(\frac{\Psi(\kappa)}{\nu} + \ln\frac{\Gamma(\kappa)}{n\Gamma(\kappa+1/\nu)}\right) & m = 1 \\ \Psi_d(m-1,n) + \Psi(m-1,\kappa)\left(\frac{d}{\nu}\right)^m & m > 1 \end{cases} \tag{6.30}$$

In principle, the parameters n, κ and ν of $CG\Gamma_P$ can be estimated by combining a system of equations from (6.29), however, it turns out to be a mathematic difficulty unless the parameter n has been estimated before. For estimating the parameter n, many approaches have been proposed in [11, 24, 26, 27]. A common method to estimate n for the whole image is by using homogeneous areas [11] which is assumed to satisfy the Wishart law. So n can be estimated from (6.26) using homogeneous regions.

For the monostatic SAR that $d = 3$, and hence, the second and third MLCs of $\mathbf{Z}^{(n)}$ can be given as

$$\tilde{\kappa}_2\{\mathbf{z}\} = \Psi_3(1,n) + 3^2(\Psi(1,\kappa)/\nu^2) \tag{6.31}$$

$$\tilde{\kappa}_3\{\mathbf{z}\} = \Psi_3(2,n) + 3^3(\Psi(2,\kappa)/\nu^3) \tag{6.32}$$

Then, we have

$$g(\kappa) = \frac{\Psi^3(1,\kappa)}{\Psi^2(2,\kappa)} = \frac{(\tilde{\kappa}_2\{\mathbf{z}\} - \Psi_3(1,n))^3}{(\tilde{\kappa}_3\{\mathbf{z}\} - \Psi_3(2,n))^2}. \tag{6.33}$$

It has been proved that $\lim_{\kappa\to 0}\Psi^3(1,\kappa)/\Psi^2(2,\kappa) = 0.25$ with the facts that $\Psi(1,\kappa) \approx \kappa^{-2}$ and $\Psi(2,\kappa) \approx -2\kappa^{-3}$ as $\kappa \to 0$ [19]. The derivative of $g(\kappa)$, denoted by $g'(\kappa)$, is deduced as

$$g'(\kappa) = \frac{\Psi^2(1,\kappa)}{\Psi^3(2,\kappa)}\left[3\Psi^2(2,\kappa) - 2\Psi(3,\kappa)\Psi(1,\kappa)\right], \tag{6.34}$$

and $g'(\kappa) > 0$ for knowing that $\Psi(2,\kappa) < 0$ and the following inequation [28]

$$\frac{n-1}{n} < \frac{\Psi^2(n,\kappa)}{\Psi(n+1,\kappa)\Psi(n-1,\kappa)} < \frac{n}{n+1}, n = 2, 3, \ldots \tag{6.35}$$

So the function $g(\kappa)$ is strictly monotonically increasing for $\kappa > 0$, which allows estimating κ with a simple numerical solution such as the bisection method.

According to (6.35) and the fact that $\Psi(2,\kappa) < 0$, it can be seen that the signs of ν and $(\tilde{\kappa}_3\{\mathbf{z}\} - \Psi_3(2,n)$ are opposite. By substituting the estimated parameter $\hat{\kappa}$ into (6.31), we have ν estimated as

$$\hat{\nu} = -\mathrm{sgn}(\tilde{\kappa}_3\{\mathbf{z}\} - \Psi_3(2,n)) \cdot 3\sqrt{\frac{\Psi(1,\hat{\kappa})}{\tilde{\kappa}_2\{\mathbf{z}\} - \Psi_3(1,n)}} \tag{6.36}$$

where $\mathrm{sgn}(\cdot)$ is the sign function.

In practice, the MLCs would be replaced by the empirical MLCs from the observed data set $\{\mathbf{Z}_i^{(n)} | i \in [1, N]\}$. The first three order empirical MLCs are computed by

$$\hat{\tilde{\kappa}}_1\{\mathbf{z}\} = \frac{1}{N}\sum_{i=1}^{N} \ln\left|\mathbf{Z}_i^{(n)}\right|, \tag{6.37}$$

$$\hat{\tilde{\kappa}}_2\{\mathbf{z}\} = \frac{1}{N}\sum_{i=1}^{N} (\ln\left|\mathbf{Z}_i^{(n)}\right| - \hat{\tilde{\kappa}}_1\{\mathbf{z}\})^2, \tag{6.38}$$

$$\hat{\tilde{\kappa}}_3\{\mathbf{z}\} = \frac{1}{N}\sum_{i=1}^{N} (\ln\left|\mathbf{Z}_i^{(n)}\right| - \hat{\tilde{\kappa}}_1\{\mathbf{z}\})^3. \tag{6.39}$$

6.4 Experimental Results and Discussions

6.4.1 Experimental Data and Evaluation Criteria

NASA/JPL AIRSAR L-band data of San Francisco is considered in this study, with its spatial resolution 10 m × 10 m, nominal number of looks 4 and dimension of 700 × 600 pixels. The fake-colored image (R: HH + VV, G: HH − VV, B: 2HV) of this scene is shown in Fig. 6.4. To assess the proposed model, four types of terrain indicated by the rectangles and numbered 1–4 in Fig. 6.4 are employed here. These four areas are the ocean area, vegetation class, mountain area and man-made urban area respectively, and hence provide a wide variety of scattering mechanism.

For quantitative assessing the goodness-of-fitting, two evaluation criteria, the mean absolute error (*MAE*) and Kolmogorov-Smirmov (*KS*) distance, are introduced and defined as follows.

(1) *MAE*: Given N samples $u_i (1 \le i \le N)$, the theoretical PDF $f(u_i)$, and the estimated PDF $q(u_i)$, the MAE can be defined as

$$MAE = \frac{1}{N}\sum_{i=1}^{N} |f(u_i) - q(u_i)| \tag{6.40}$$

Fig. 6.4 Fake-colored image of the San Fransisco

6.4.2 Modeling Result

The validation of the proposed model is performed by using the marginal intensity distributions for each channel as it was done by Lee [10] and Freitas [11]. For comparison, the classical distributions, Wishart, $\mathcal{K}_P$ and $\mathcal{G}_P^0$ distribution are included. All the parameters of each distribution are estimated by the MoMLC and listed in Table 6.1, where $\hat{n}$ is estimated with the data of ocean area numbered 1 that assumed to be homogeneous.

It can be seen from Table 6.1 that:

(1) the estimated parameter $\hat{n}$ representing the ENL is 4.1191, which is close to the nominal ENL 4 and shows the validation of the estimator for ENL mentioned above. We applied the estimated value of $\hat{n}$ to all the other distributions considered;
(2) the parameter $\hat{\kappa}$ of $\mathcal{G}_P^0$ distribution is named as roughness parameter in [11], of which the smaller absolute value means more heterogeneous area. The value of area 4 is between that of area 2 and 3, which is near to our common sense;
(3) the parameter $\hat{\kappa}$ of $\mathcal{G}_P^0$ and that of $\mathcal{K}_P$ own the same absolute value but opposite sign arising from the similarity of their second MLC equations.

The fitted distributions to the data numbered from 1–4 are shown in Fig. 6.5, 6.6, 6.7 and 6.8 respectively, and the corresponding *MAE* and *KS* values are listed in Table 6.2.

As can be seen from Fig. 6.5, 6.6, 6.7 and 6.8 and Table 6.2, we can find that:

Table 6.1 Estimated parameters of three considered areas in Fig. 6.4

Tab		1	2	3	4
Type		Ocean	Vegetation	Mountain	Urban
Sample size		10,611	9301	12,231	15,691
$\hat{\mathbf{C}}$	$\hat{C}_{11}$	0.0078	0.0434	0.0937	0.1969
	$\hat{C}_{22}$	0.0004	0.0187	0.0200	0.0280
	$\hat{C}_{33}$	0.0261	0.0607	0.1134	0.1738
	$\hat{C}_{12}$	0.0004 −0.0007 i	0.0026 −0.0015 i	0.0073 −0.0042 i	0.0362 −0.0014 i
	$\hat{C}_{13}$	0.0134 +0.0018 i	0.0068 +0.0033 i	0.0537 +0.0044 i	−0.0437 −0.0045 i
	$\hat{C}_{23}$	0.0006 +0.0014 i	0.0009 +0.0020 i	0.0071 +0.0015 i	−0.0127 +0.0058 i
Wishart	$\hat{n}$	4.1191			
$\mathcal{K}_P$	$\hat{\kappa}$	10.2783	2.7084	1.3683	2.1103
$\mathcal{G}_P^0$	$\hat{\kappa}$	−10.2783	−2.7084	−1.3683	−2.1103
$\mathcal{C}\mathcal{G}\Gamma_P$	$\hat{\kappa}$	2.3289	5.5965	2.3771	1.1591
	$\hat{\nu}$	−2.2873	0.6625	−0.7020	−1.4852

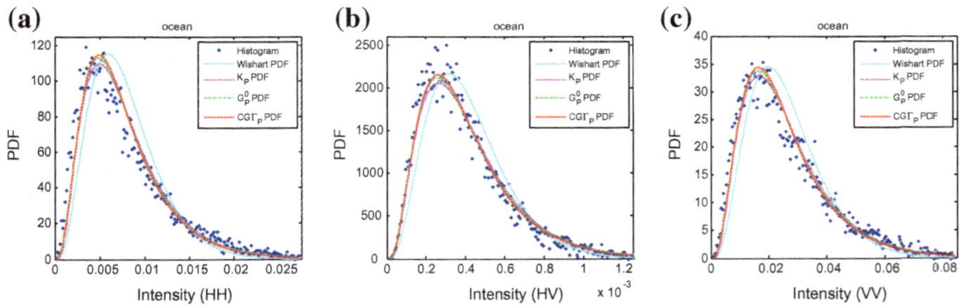

Fig. 6.5 Fitted distribution to the ocean data. **a** HH channel, **b** HV channel, **c** VV channel

(1) The Wishart distribution fails to describe the areas numbered from 2 to 4, and can only fit to the homogeneous ocean area. This is because the limit central theorem that the Wishart model based on is relatively applicable only for the ocean area of the considered data. However, it is the worst one compared to the other three models.

(2) $\mathcal{K}_P$ distribution provides a well fit for both the ocean and vegetation areas but fails to the extremely heterogeneous mountain area and man-made urban area. The results is similar to that in literature [11].

(3) Both $\mathcal{G}_P^0$ and $\mathcal{C}\mathcal{G}\Gamma_P$ distribution can explain all the considered areas well, whilst $\mathcal{C}\mathcal{G}\Gamma_P$ is better than $\mathcal{G}_P^0$ especially for the mountain area where $\mathcal{G}_P^0$ has an obvi-

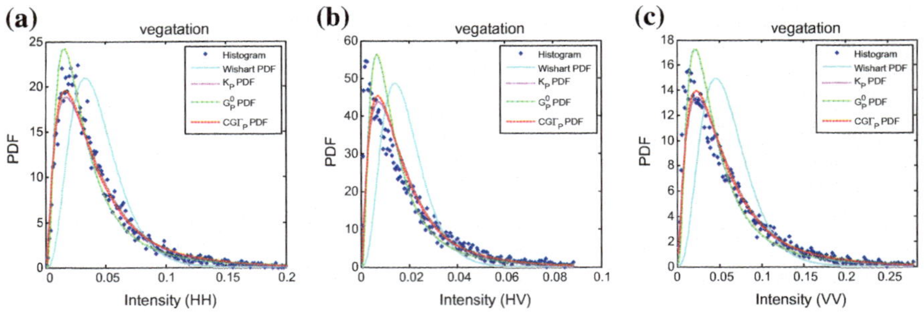

Fig. 6.6 Fitted distribution to the vegetation data. **a** HH channel, **b** HV channel, **c** VV channel

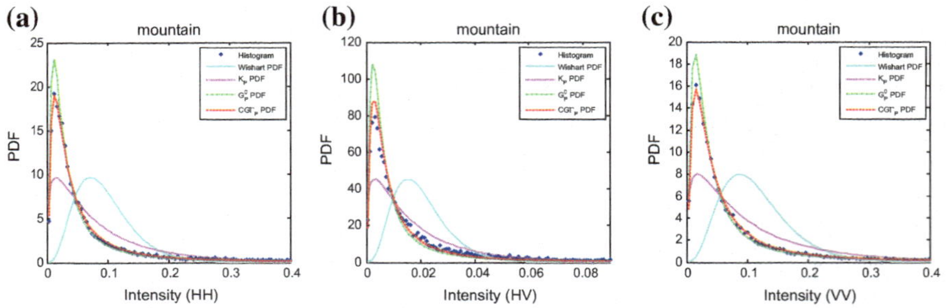

Fig. 6.7 Fitted distribution to the mountain data. **a** HH channel, **b** HV channel, **c** VV channel

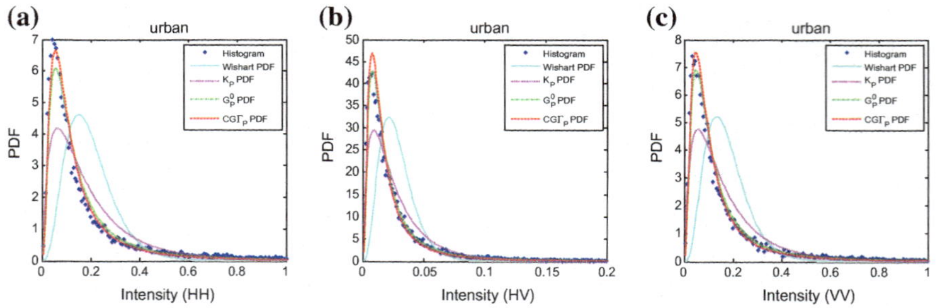

Fig. 6.8 Fitted distribution to the urban data. **a** HH channel, **b** HV channel, **c** VV channel

ously higher peak than the histogram of the real data and $CG\Gamma_P$ provides a well fit. The same conclusion can be drawn by the quantitative evaluation criteria that $CG\Gamma_P$ owns the lower value of *MAE* and *KS* for most situations.

The KummerU distribution should also be employed for compare here, however, we found that the solution of MoMLC for KummerU distribution (see literature [23]) is unstable that may have different results with different initial iterative values when the numerical method is applied. It's because that the surface of the third MLC $\tilde{\kappa}_3$ versus the two unknown parameters is not convex.

Table 6.2 Values of Max and KS corresponding to the fitting results in Fig. 6.9

Tap			1	2	3	4
MAE	HH	*Wishart*	10.9779	2.2923	1.1961	0.6100
		$\mathcal{K}_P$	5.4455	0.6298	0.5049	0.2627
		$\mathcal{G}_P^0$	5.3525	0.8095	0.1765	0.1207
		$CG\Gamma_P$	5.5124	0.6183	0.0978	0.1061
	HV	*Wishart*	297.3059	5.0584	6.6891	3.7768
		$\mathcal{K}_P$	97.3637	2.6789	1.9094	1.2482
		$\mathcal{G}_P^0$	178.1651	2.0462	0.5469	0.5863
		$CG\Gamma_P$	172.6286	1.4399	0.7988	0.5823
	VV	*Wishart*	4.1208	1.5909	0.8391	0.6438
		$\mathcal{K}_P$	1.1294	0.7986	0.1872	0.2266
		$\mathcal{G}_P^0$	2.2193	0.6324	0.1937	0.1135
		$CG\Gamma_P$	2.1738	0.4938	0.0966	0.1055
KS	HH	*Wishart*	0.2985	0.4616	1.0531	0.8732
		$\mathcal{K}_P$	0.1481	0.1268	0.4445	0.3760
		$\mathcal{G}_P^0$	0.1456	0.1630	0.1554	0.1727
		$CG\Gamma_P$	0.1499	0.1245	0.0861	0.1519
	HV	*Wishart*	0.3704	0.4474	1.1119	0.8304
		$\mathcal{K}_P$	0.1213	0.2369	0.3174	0.2744
		$\mathcal{G}_P^0$	0.2220	0.1810	0.0909	0.1289
		$CG\Gamma_P$	0.2151	0.1273	0.1328	0.1280
	VV	*Wishart*	0.3456	0.4459	1.0035	0.8519
		$\mathcal{K}_P$	0.0947	0.2238	0.2239	0.2999
		$\mathcal{G}_P^0$	0.1861	0.1773	0.2317	0.1502
		$CG\Gamma_P$	0.1823	0.1384	0.1155	0.1396

6.4.3 Discussions

Under the multiplicative model framework, this chapter proposed a novel distribution for multilook polarimetric covariance matrix by modeling the backscatter component with the generalized gamma distribution. This distribution covers a larger range than the KummerU distribution in the famous $\tilde{\kappa}_3 \sim \tilde{\kappa}_2$ diagram and can be seen as an extension of the classic Wishart, $\mathcal{K}_p$ and $\mathcal{G}_P^0$ distributions. Furthermore, we derive a corresponding parameter estimator based on the method of matrix log-cumulants.

The experimental results on actual images of different scattering mechanism types validate the flexibility of the proposed model and the effectiveness of the estimator. It would be attractive to apply this model in the PolSAR image processing, like classification, segmentation and target detection, etc.

6.5 Ship Detection in High-Resolution Dual Polarization SAR Amplitude Images

Ship detection in high-resolution SAR images has become an increasing interest during past several years [29]. It plays an important role for various potential applications like marine surveillance, vessel traffic control, military service, etc. [29–31]. Additionally, with the development of sensor techniques, the advanced polarimetric SAR systems have also been validated that more complete description of target scattering behavior could be provided than traditional single channel SAR systems, such as HH, HV and VV SAR, which is not sufficient for ship detection [30], [32]. Since better discriminating target signal from the surrounding clutter can be obtained by polarimetric SAR systems, ship detection in multi polarimetric or dual polarimetric (dual-pol) SAR images are receiving intensive attentions in present.

Nowadays, some famous contemporary satellite SAR systems, e.g., TerraSAR-X, COSMO-SkyMed, as well as RadarSat-2, support dual-pol acquisition mode [33]. However, the available data is, mostly, the amplitude data and not the complex-valued data [34]. It is an important data modality because the image products provided by many satellite SAR systems, for instance, ERS. Thus, we mainly focus on ship detection using only dual-pol SAR amplitude images in this chapter.

A wide variety of methods have been proposed for detecting ships in polarimetric SAR images. The basic idea is to reduce three channels of polarimetric SAR data to a single decision criterion [35]. Some popular detectors [35], including the optimal polarimetric detector (OPD), the polarimetric whitening filter (PWF), the span detector, the power maximization synthesis (PMS) detector, and more recently, the identity-likelihood-ratio-test (ILRT), have been developed and shown they could perform in a way that targets are more easily discriminated from clutter. Unfortunately, other detectors mentioned-above except the span detector are very difficult to implement due to the missing of phase information when the only amplitude is considered [35].

The span detector is a widely used processor which is a non-coherent sum of all polarimetric channels and only makes use of image intensities or amplitudes. It has also been proven that this detector can acquire a better detection performance than HH, HV, or VV individually. Nevertheless, the expected adaptive detection threshold is impossible when facing complex clutter background by this detector, which limits the applications of this detector.

Our goal is to present a powerful detector for ship detection in high-resolution dual-pol SAR amplitude images. On one hand, this detector can improve the signal-to-clutter ratio (SCR) to enhance the moving targets or restrain the clutter. Meanwhile, we wish that a flexible and adaptive const false alarm rate (CFAR) threshold could be derived from this detector. Under this consideration, this chapter proposed a novel detector similarly with the span detector, simply called the product of multi-look amplitudes (PMA) detector. Simultaneously, the PMA's statistical model and the parameter estimation method as well as the corresponding CFAR threshold are also presented.

6.5.1 Dual-Pol SAR Data Description

6.5.1.1 Polarimetric Covariance Matrix for Dual-Pol Case

For dual-pol case, the single look scattering vector can be simplified to

$$\mathbf{u} = \begin{bmatrix} S_1 & S_2 \end{bmatrix}^{\mathrm{T}} \tag{6.41}$$

Herein, for convenience, we use S_1 or S_2 to indicate one of the scattering elements S_{hh}, S_{hv} and S_{vv} in any order, as well as $S_1 \neq S_2$. Additionally, to reduce the influence of speckle, SAR data are often multi-look averaged. As the polarimetric information can also be represented by a covariance matrix, the n-look sample covariance matrix is defined as [36, 37]

$$\mathbf{R} = \frac{1}{n}\sum_{k=1}^{n} \mathbf{u}(k)\mathbf{u}(k)^{H} = \frac{1}{n}\sum_{k=1}^{n} \begin{bmatrix} |S_1(k)|^2 & S_1(k)S_2(k)^* \\ S_1(k)^*S_2(k) & |S_2(k)|^2 \end{bmatrix} \tag{6.42}$$

where the superscript * means complex conjugate and H represents conjugate complex transpose, n is the number of looks and $\mathbf{u}(k) = [S_1(k), S_2(k)]^T$ is the k-th single-look image. Assuming statistical ergodicity and constant RCS background, the random matrix $\mathbf{R}$, known as the complex Wishart distribution [38], is with probability density

$$p_{\mathbf{R}}(\mathbf{R}) = \frac{n^{2n}\det(\mathbf{R})^{n-2}\exp\left[-n\mathrm{Tr}(\mathbf{C}^{-1}\mathbf{R})\right]}{\pi\Gamma(n)\Gamma(n-1)\det(\mathbf{C})^{n}} \tag{6.43}$$

where $\Gamma(\cdot)$ is the gamma function, and $\mathrm{Tr}(\cdot)$ indicates the matrix trace. The symbol $\det(\cdot)$ denotes the determinant operator and the covariance matrix is 2×2 complex, Hermitian, written as

$$\mathbf{C} = \mathrm{E}[\mathbf{u}\mathbf{u}^{\mathrm{H}}] = \begin{bmatrix} E(|S_1|^2) & \sqrt{E(|S_1|^2)E(|S_2|^2)}\rho e^{j\theta} \\ \sqrt{E(|S_1|^2)E(|S_2|^2)}\rho e^{-j\theta} & E(|S_2|^2) \end{bmatrix} \tag{6.44}$$

where $\rho e^{j\theta}$ is the complex correlation coefficient of two components.

6.5.1.2 The Joint Distribution of Two Multi-look Intensities from Different Polarimetric Channels

When only incomplete polarimetric data are available, for instance, the amplitude or intensity of copolarized components (i.e., *HH* and *VV*), the joint distribution of intensity or amplitude from two correlated polarimetric channels is of important for constructing signal processing algorithms, such as detection, classification, and so on, in this case. Lee et al. [9] have derived the PDF of joint normalized multi-look intensities by integrating (6.43) with respect to the two off-diagonal elements, which is modeled as

$$p_{R_1,R_2}(R_1, R_2) = \frac{n^{n+1}(R_1R_2)^{\frac{n-1}{2}}\exp\left(-\frac{n(R_1+R_2)}{1-\rho^2}\right)}{\Gamma(n)(1-\rho^2)\rho^{n-1}} I_{n-1}\left(2n\sqrt{R_1R_2}\frac{\rho}{1-\rho^2}\right),$$
$$R_1, R_2, n > 0, \textit{ and } 0 < \rho < 1 \tag{6.45}$$

where $I_{n-1}(\cdot)$ is the first type modified Bessel function of order $n-1$, R_i, $i = 1, 2$ represents normalized multi-look intensity of i-th polarimetric channel with the expression $R_i = \frac{1}{n}\sum_{k=1}^{n}\frac{|S_i(k)|^2}{E(|S_i|^2)}$.

Furthermore, to facilitate the posterior derivation, by the transforms with the following forms:

$$B_1 = nR_1$$
$$B_2 = nR_2 \tag{6.46}$$

One can obtain the joint density of B_1 and B_2 as (see [36] in detailed)

$$p_{B_1,B_2}(B_1, B_2) = \frac{(B_1B_2)^{\frac{n-1}{2}}\exp\left(-\frac{B_1+B_2}{1-\rho^2}\right)}{\Gamma(n)(1-\rho^2)\rho^{n-1}} I_{n-1}\left(2\sqrt{B_1B_2}\frac{\rho}{1-\rho^2}\right),$$
$$B_1, B_2, n > 0, \textit{ and } 0 < \rho < 1 \tag{6.47}$$

6.5.2 The PMA Detector

6.5.2.1 PMA Detector

In a single channel SAR image, it is usually assumed that strong backscattering comes from targets. A target point will be lost when the backscattering amplitude or intensity from the target is not large enough compared with the clutter background, generally, due to low signal-to-clutter ratio (SCR). In other words, SCR is an essential factor influencing the detection performance when only amplitude or intensity data are available. Thus, the principle designing a good detector should naturally enable SCR to be improved, i.e., enhancing target and restraining clutter.

As we known, the span detector is a widely used processor which is a non-coherent sum of all polarimetric channels and only makes use of image intensities. For dual-pol multilook case, the span is given by [35]

$$span = \frac{1}{n}\sum_{k=1}^{n}|S_1(k)|^2 + \frac{1}{n}\sum_{k=1}^{n}|S_2|^2 \tag{6.48}$$

This detector can be regarded as the synthetic power of all channels. Consequently, some investigations have shown that a lower noise level and a higher SCR can be obtained by this detector than HH, HV, or VV individually. This conclusion implies that the way of synthetic power can arrive at the purpose of improving SCR, so that the targets can be more easily discriminated from the clutter, compared with that only arbitrary single channel information is used. Meanwhile, it is also very hard to adaptively give a proper detection threshold by the span detector due to the unknown knowledge of the corresponding statistics. Motivated by these considerations, we construct a detector in this chapter by means of another synthetic power, i.e., the product of multi-look amplitudes from two polarimetric channels, for convenience, called PMA detector which can be defined as:

$$\xi = \sigma_1\sigma_2\sqrt{R_1R_2} = \sigma_1\sigma_2\frac{\sqrt{B_1B_2}}{n} \tag{6.49}$$

where $\sigma_i = E\left(|S_i|^2\right)$. From an intuitive understanding, in the variable ξ domain, for the targets, their amplitudes of two polarimetric channels are both larger than surrounding clutter, which results in a much faster cumulative speed of power for targets verse clutter by multiplying the first image amplitude by the second image one, and hence targets' values of ξ are much more prominent and target signal can be enhanced.

6.5.2.2 Statistics of the Proposed Detector

Letting $\varsigma = \frac{\xi}{\sigma_1\sigma_2}$, ς indicates the product of normalized multi-look amplitudes from two polarimetric channels. As we can see from (6.49), it allows us to begin derivation of ξ with the PDF of two random variables' product B_1B_2. We let

$$\eta = B_1B_2 \tag{6.50}$$

As given in Appendix 6.2 based on (6.47), the PDF of η is derived as

$$p_\eta(\eta) = \frac{2\eta^{\frac{n-1}{2}}}{\Gamma(n)(1-\rho^2)\rho^{n-1}} I_{n-1}\left(2\sqrt{\eta}\frac{\rho}{1-\rho^2}\right)K_0\left(\frac{2\sqrt{\eta}}{1-\rho^2}\right),$$
$$\eta, n > 0,\ and\ 0 < \rho < 1 \tag{6.51}$$

where $K_0(\cdot)$ indicates the second type modified Bessel function with order 0.

According to (6.49) and (6.51), it is clear that $\varsigma = \frac{\sqrt{\eta}}{n}$. Consequently, the PDF, denoted simply as $p_\varsigma(\varsigma)$, about the product of normalized multi-look amplitudes from two polarimetric channels can be characterized by

$$p_\varsigma(\varsigma) = \frac{4\varsigma^n n^{n+1}}{\Gamma(n)\left(1-\rho^2\right)\rho^{n-1}} I_{n-1}\left(\frac{2n\rho}{1-\rho^2}\varsigma\right)K_0\left(\frac{2n}{1-\rho^2}\varsigma\right),$$
$$\varsigma, n > 0,\ and\ 0 < \rho < 1 \tag{6.52}$$

Furthermore, resorting to the asymptotic formulas of the modified Bessel functions, the resulting PDF of ξ is (the detailed derivations are included in Appendix 6.3)

$$p_\xi(\xi) = \frac{\beta^n}{\Gamma(n)}\xi^{n-1}\exp(-\beta\xi), \quad \xi, n, \beta > 0 \tag{6.53}$$

with the parameter $\beta = \frac{2n}{(1+\rho)\sigma_1\sigma_2}$. Its moments of s order turn out to be

$$E(\xi^s) = \frac{\Gamma(s+n)}{\Gamma(n)\beta^s} \tag{6.54}$$

6.5.2.3 Performance Analysis of PMA Detector

In this part, we further discuss the performance of PMA detector. In other words, we would like to theoretically give how to improve the SCR by the proposed detector. For the high-resolution polarimetric SAR radar, a target is modeled as deterministic, individually resolved unitary scatterer with extended pixels in image. For convenience of discussion, when we consider a point target (the extended case is similar) situation, the SCR of the i-th polarimetric channel image can be defined in dB style by

$$SCR^{(i)} = 20\log_{10}\left(\frac{X_T^{(i)}}{\sigma_i}\right) \tag{6.55}$$

where $X_T^{(i)}$ indicates the target amplitude in the i-th polarimetric channel image, σ_i are the mean value of clutter pixel's amplitudes in the same image.

Under the condition that the target can be detected, $\frac{X_T^{(i)}}{\sigma_i} > 1$. Then the SCR in the product image of multi-look amplitudes from two polarimetric channels is

$$SCR_P = 20\log_{10}\left(\frac{X_T^{(1)}X_T^{(2)}}{E\left(Y_C^{(1)}Y_C^{(2)}\right)}\right) \tag{6.56}$$

Based on Sect. 6.5.2.2, $Y_C^{(1)}Y_C^{(2)}$ is the clutter amplitude product and obeys the density shown in (6.53), which leads to $E(Y_C^{(1)}Y_C^{(2)}) = \frac{(1+\rho)\sigma_1\sigma_2}{2}$ from (6.54). Therefore, (6.56) is further given by

$$SCR_P = 20\log_{10}\left(\frac{2X_T^{(1)}X_T^{(2)}}{(1+\rho)\sigma_1\sigma_2}\right) \tag{6.57}$$

Because $\rho < 1$, $\frac{2}{1+\rho} > 1$. Additionally, due to that $\frac{X_T^{(1)}}{\sigma_1} > 1$ and $\frac{X_T^{(2)}}{\sigma_2} > 1$, $\frac{2X_T^{(1)}X_T^{(2)}}{(1+\rho)\sigma_1\sigma_2} > \frac{X_T^{(i)}}{\sigma_i}$. Considering the monotonously increasing characteristic of logarithm function, $SCR_P > SCR^{(i)}$. The above analysis testifies the PMA metric has a greater SCR value than arbitrary single channel, since two channel images are correlated in dual-pol SAR system.

6.5.3 *The CFAR Algorithm of PMA Detector*

6.5.3.1 The Statistics of PMA Detector in in-Homogeneous Clutter

As we can see from the derivation procedure and the analytical form shown in (6.53), it suffers by the following limitation in practice when (6.53) is used. Clearly, the derivation of (6.53) originally begins with a multivariate complex Wishart distribution as a statistical model for the statistics of the covariance matrix, yet this model is validated to be suited for a constant RCS background since the hypothesis that no dominant scatters exist in any resolution cell of each channel returns has been made. This means a large deviation will be generated in fitting the circumstance of in-homogeneous clutter by (6.53). Fortunately, from lots of literatures, the well-known multiplicative model has been found to be a feasible tool to acquire the corresponding PDFs of heterogeneous or extremely heterogeneous terrains, like forest and sea clutter as well as urban region, by separating RCS component (also named texture component) and speckle component and characterizing them respectively. For dual-pol SAR amplitude data, based on the multiplicative model, the product of multi-look amplitudes from two polarimetric channels can be denoted as

$$\zeta = \sqrt{\frac{1}{n}\sum_{k=1}^{n}|A_1X_1(k)|^2\frac{1}{n}\sum_{k=1}^{n}|A_2X_2(k)|^2} = \sqrt{A_1^2A_2^2\frac{1}{n}\sum_{k=1}^{n}|X_1(k)|^2\frac{1}{n}\sum_{k=1}^{n}|X_2(k)|^2} = A_1A_2\xi \tag{6.58}$$

where A_i represents the backscattering RCS amplitude component of i th receiving polarimetric channel and $X_i(k) = \frac{|S_i(k)|^2}{E(|S_i|^2)}$ which indicates speckle component with $X_i \sim \mathcal{N}^{\mathbb{C}}(0, 1)$, $\mathcal{N}^{\mathbb{C}}(0, 1)$ denotes the complex normal distribution with expectation 0 and variance 1. An usual assumption is that the energy of two-channel is balanced, i.e., $A_1 = A_2 = A$, which results in

$$\zeta = A^2\xi \tag{6.59}$$

Additionally, a reciprocal of the square root of a Gamma distribution [13] has been widely suggested to describe the RCS fluctuation of clutter amplitude for heterogeneous and extremely heterogeneous regions. Consequently, the corresponding intensity, denoted simply as $W = A^2$ and is with the PDF, known as a reciprocal of a Gamma distribution, given by [13]

$$p_W(w) = \frac{\gamma^{-\alpha}}{\Gamma(-\alpha)}w^{\alpha-1}\exp\left(-\frac{\gamma}{w}\right), \quad -\alpha, \gamma > 0 \tag{6.60}$$

where α ($-\alpha \in (0, \infty)$) is a shape parameter, which essentially reflects the degrees of homogeneity for processed areas. γ is a scale parameter related to the mean energy of processed areas. In term of the probability density formula of the multiplication of two variables, $p_\zeta(\zeta) = \int_0^\infty \frac{1}{t}p_W(t)p_\xi\left(\frac{\zeta}{t}\right)dt$, as well as the integral formula $\int_0^\infty x^m\exp(-\beta x^n)dx = \frac{\Gamma(\gamma)}{n\beta^\gamma}$, $\gamma = \frac{m+1}{n}$, $[\mathrm{Re}\,\beta > 0,\ \ \mathrm{Re}\,m > 0,\ \ \mathrm{Re}\,n > 0]$, one obtains the mathematic expression by combining (6.53), (6.59) and (6.60) as

$$p_\zeta(\zeta) = \frac{\sigma^n}{B(n, -\alpha)}\frac{\zeta^{n-1}}{(1+\sigma\zeta)^{n-\alpha}}, \quad \sigma, n, -\alpha, \zeta > 0 \tag{6.61}$$

where $\sigma = \frac{\beta}{\gamma}$, $B(\cdot, \cdot)$ is the beta function. Obviously, as we can easily see from (6.61), the product of multi-look amplitudes from two polarimetric channels, i.e., the proposed detector in this chapter, is proven to obey a well-known $\mathcal{G}^0$ intensity distribution with three distinct parameters σ, n and α, when applying to the modeling of sea background. This means both single channel intensity and the product of dual channel amplitude are perfectly consolidate into the same distribution, their differentiation only lies on the different parameter values and they can be regarded as the generalization of single-channel SAR images. The estimates $\hat{\alpha}$, $\hat{\sigma}$ and $\hat{n}$ corresponded respectively to the parameters α, σ and n can be easily obtained with the help of numerical calculation based on the MoLC [38] as

$$\begin{cases} \Psi(\hat{n}) - \Psi(-\hat{\alpha}) - \ln(\hat{\sigma}) = \frac{1}{N}\sum\limits_{i=1}^{N}[\ln(x_i)] = \hat{\tilde{c}}_1 \\ \Psi(1,\hat{n}) + \Psi(1,-\hat{\alpha}) = \frac{1}{N}\sum\limits_{i=1}^{N}\left[\left(\ln(x_i) - \hat{\tilde{c}}_1\right)^2\right] \\ \Psi(2,\hat{n}) - \Psi(2,-\hat{\alpha}) = \frac{1}{N}\sum\limits_{i=1}^{N}\left[\left(\ln(x_i) - \hat{\tilde{c}}_1\right)^3\right] \end{cases} \tag{6.62}$$

where $\Psi(\cdot)$ represents the digamma function (i.e., the logarithmic derivative of the Gamma function), $\Psi(r,\cdot)$ is the r-th order polygamma function (i.e., the kth order derivative of the digamma function), and $\{x_i\}$, $i \in [1, N]$ is a given sample set.

6.5.3.2 CFAR Detection

Given the density shown in (6.61), its CDF is written as [13]

$$F_\zeta(x) = \frac{\sigma^n x^n}{nB(n,-\alpha)}\,{}_2F_1(n-\alpha, n; n+1; -\sigma x) \tag{6.63}$$

where ${}_2F_1(\cdot\,,\,\cdot;\,\cdot;\,\cdot)$ is the Gauss hypergeometric function. For a given value of the false alarm probability, denoted by P_{fa}, the corresponding CFAR threshold T for the distribution shown in (6.61) can be obtained from

$$1 - P_{fa} = F_\zeta(T) = \frac{\sigma^n T^n}{nB(n,-\alpha)}\,{}_2F_1(n-\alpha, n; n+1; -\sigma T) \tag{6.64}$$

Considering $F_\zeta(T)$ is strictly monotonously increasing, the threshold T can be accurately calculated via the numerical solution or a simple bisection method.

6.5.4 Experimental Results and Analysis

The test dual-pol SAR amplitude data used in this study are a large TerraSAR-X StripMap-mode geocoded scene over Nanjing of China, acquired with high-resolution 6 m × 6 m (azimuth × range) and HH-polarization and VV-polarization. Figure 6.9a provides a fake-color image of this scene. The horizontal and vertical axes are the directions of azimuth and range, respectively. Meanwhile, in order to make a visible comparison, Fig. 6.9b gives an optical (ground truth) remote sensing photograph of the test site from SPOT5 satellite.

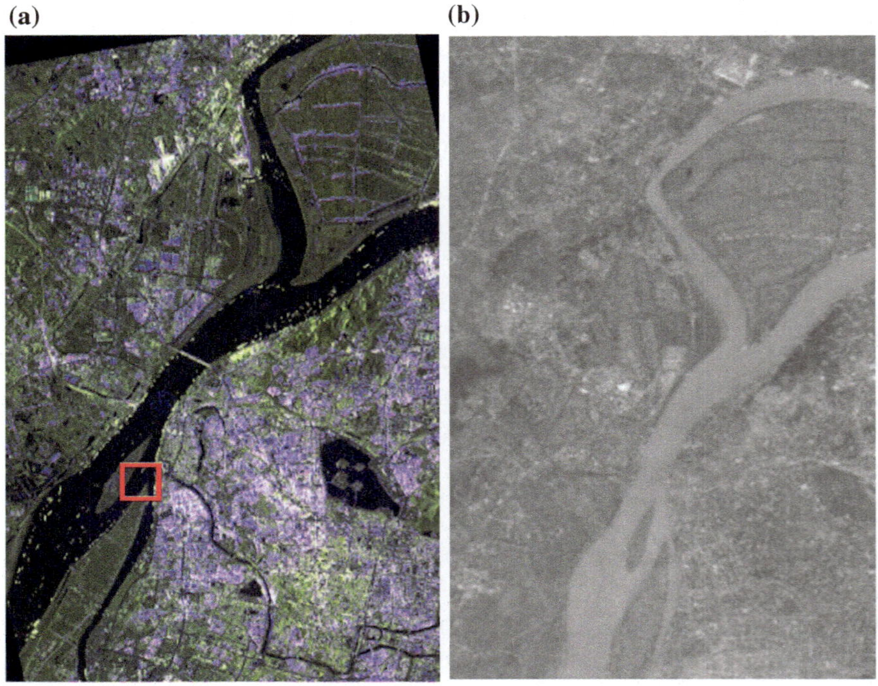

Fig. 6.9 The image of Nanjing: **a** TerraSAR-X image; **b** SPOT5 optical image

The red rectangle box region shown in Fig. 6.9a consisting of several ships and sea clutter is our investigating area. The product image of multi-look amplitudes from two polarimetric channels is shown as Fig. 6.10a. Figure 6.10b shows the fitting results of the distribution in (6.61) for the product image of the area indicated in the rectangle box in Fig. 6.10a. The parameters n, α and σ are estimated to $\hat{n} = 100.5744$, $\hat{\alpha} = -1.7851$ and $\hat{\sigma} = 120.7748$, respectively. As shown in Fig. 6.10b, it is clear that the fitting result agrees well with the theoretical distribution.

Furthermore, giving the theoretical false alarm probability $P_{fa} = 10^{-8}$, the detection results are shown in Fig. 6.11. It is easy to observe that all ships are detected, whilst a false alarm occurs, which proves the effectiveness of CFAR detection method based on PMA detector for ships.

6.5.5 *Experimental Results and Analysis*

Aiming at the adaptive detection of ship when only high-resolution dual-pol SAR amplitude data are available, A CFAR detecting method has been proposed in this chapter. We first design a novel PMA detector, which can improve the SCR and make

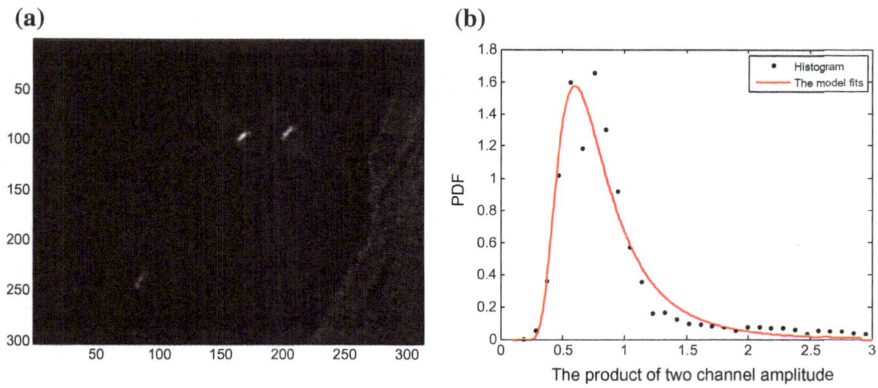

Fig. 6.10 The ship chip: **a** the product image; **b** the fitting result

Fig. 6.11 The ship detecting result

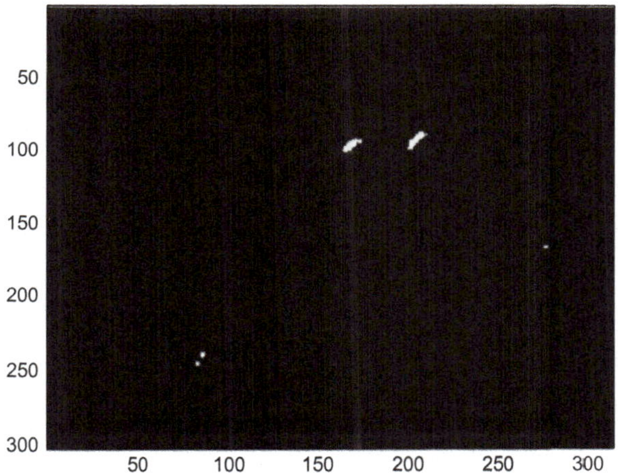

the discrimination of ship from clutter more easily. Meanwhile, under the frame of multiplicative model, based on a reciprocal of the square root of a Gamma distribution to describe the texture component of returns, the PMA detector's statistical model has been derived when facing complex sea background. Finally, the parametric estimators and the CFAR threshold have also been given analytically. The experiments performed on measured dual-pol TerraSAR-X images demonstrate the good performance of the proposed CFAR detecting method.

Appendix 6.1: The Derivation of $CG\Gamma_P$ Distribution Toward the $\mathcal{K}_P$ and $\mathcal{G}_P^0$ Distributions

Let $\nu = 1$, then we have $\xi = \Gamma(\kappa + 1)/\Gamma(\kappa) = \kappa$, and hence (6.8) can be rewritten as

$$f_{\mathbf{Z}^{(n)}}(\mathbf{z}) = \frac{n^{nd}|\mathbf{z}|^{n-d}\kappa^{\kappa}}{\Gamma_d(n)|\mathbf{C}|^n\Gamma(\kappa)} \cdot \int_0^{\infty} x^{\kappa-nd-1}\exp\left\{-\frac{nTr(\mathbf{C}^{-1}\mathbf{z})}{x} - \kappa x\right\}dx \tag{6.65}$$

By using the integral of the modified Bessel function of the second kind [39]:

$$2(a/b)^{m/2}K_m(2\sqrt{ab}) = \int_0^{\infty} x^{m-1}\exp(-a/x - bx)dx, \quad a, b > 0$$

we have

$$\begin{aligned} f_{\mathbf{Z}^{(n)}}(\mathbf{z}) &= \frac{n^{nd}|\mathbf{z}|^{n-d}\kappa^{\kappa}}{\Gamma_d(n)|\mathbf{C}|^n\Gamma(\kappa)} \cdot 2\left(nTr(\mathbf{C}^{-1}\mathbf{z})/\kappa\right)^{(\kappa-nd)/2}K_{\kappa-nd}(2\sqrt{\kappa nTr(\mathbf{C}^{-1}\mathbf{z})}) \\ &= 2\frac{|\mathbf{z}|^{n-d}(n\kappa)^{(\kappa+nd)/2}}{\Gamma_d(n)|\mathbf{C}|^n\Gamma(\kappa)} \cdot \frac{K_{\kappa-nd}(2\sqrt{\kappa nTr(\mathbf{C}^{-1}\mathbf{z})})}{(Tr(\mathbf{C}^{-1}\mathbf{z}))^{-(\kappa-nd)/2}} \end{aligned} \tag{6.66}$$

which is the $\mathcal{K}_P$ distribution proposed in [10].

Let $\nu = -1$, then we have $\xi = \Gamma(\kappa - 1)/\Gamma(\kappa) = (\kappa - 1)^{-1}$, and hence (6.8) can be rewritten as

$$\begin{aligned} f_{\mathbf{Z}^{(n)}}(\mathbf{z}) &= \frac{n^{nd}|\mathbf{z}|^{n-d}(\kappa-1)^{\kappa}}{\Gamma_d(n)|\mathbf{C}|^n\Gamma(\kappa)} \cdot \int_0^{\infty} x^{-\kappa-nd-1}\exp\left\{-\frac{nTr(\mathbf{C}^{-1}\mathbf{z}) + \kappa - 1}{x}\right\}dx \\ &= \frac{n^{nd}|\mathbf{z}|^{n-d}(\kappa-1)^{\kappa}}{\Gamma_d(n)|\mathbf{C}|^n\Gamma(\kappa)} \cdot \int_0^{\infty} t^{\kappa+nd-1}\exp\{-(nTr(\mathbf{C}^{-1}\mathbf{z}) + \kappa - 1)t\}dt \end{aligned} \tag{6.67}$$

By using the integral of the gamma function [39]:

$$\int_0^{\infty} x^{\nu-1}e^{-\mu x}dx = \Gamma(\nu)/\mu^{\nu}, \quad \mu, \nu > 0 \tag{6.68}$$

one can obtain

$$f_{\mathbf{Z}^{(n)}}(\mathbf{z}) = \frac{n^{nd}|\mathbf{z}|^{n-d}(\kappa-1)^{\kappa}}{\Gamma_d(n)|\mathbf{C}|^n\Gamma(\kappa)} \cdot \frac{\Gamma(\kappa + nd)}{(nTr(\mathbf{C}^{-1}\mathbf{z}) + \kappa - 1)^{\kappa+nd}} \tag{6.69}$$

Let $\kappa = -\alpha$, (6.69) can rewritten as

$$f_{\mathbf{Z}^{(n)}}(\mathbf{z}) \overset{\kappa=-\alpha}{=} \frac{n^{nd}|\mathbf{z}|^{n-d}\Gamma(nd-\alpha)}{\Gamma_d(n)|\mathbf{C}|^n\Gamma(-\alpha)(-\alpha-1)^{\alpha}} \cdot (nTr(\mathbf{C}^{-1}\mathbf{z})-\alpha-1)^{\alpha-nd} \tag{6.70}$$

which has the same form of $\mathcal{G}_P^0$ presented in [11].

Appendix 6.2: The Derivation of the Distribution of B_1B_2

Via (6.50), we get the following joint PDF:

$$\begin{aligned} p_{B_1,\eta}(B_1,\eta) &= \frac{1}{B_1}p_{B_1,B_2}\left(B_1,\frac{\eta}{B_1}\right) \\ &= \frac{\eta^{\frac{n-1}{2}}}{\Gamma(n)(1-\rho^2)\rho^{n-1}}I_{n-1}\left(2\sqrt{\eta}\frac{\rho}{1-\rho^2}\right)\frac{1}{B_1}\exp\left(-\frac{B_1}{1-\rho^2}-\frac{\eta}{1-\rho^2}\frac{1}{B_1}\right) \end{aligned} \tag{6.71}$$

Then, the marginal density of η is

$$\begin{aligned} p_\eta(\eta) &= \int_0^\infty p_{B_1,\eta}(B_1,\eta)dB_1 \\ &= \frac{\eta^{\frac{n-1}{2}}}{\Gamma(n)(1-\rho^2)\rho^{n-1}}I_{n-1}\left(2\sqrt{\eta}\frac{\rho}{1-\rho^2}\right)\int_0^\infty \frac{1}{B_1}\exp\left(-\frac{1}{1-\rho^2}B_1-\frac{\eta}{1-\rho^2}\frac{1}{B_1}\right)dB_1 \end{aligned} \tag{6.72}$$

By a variable change $x = \frac{2B_1}{1-\rho^2}$, the integral component of the right part in (6.72) can be further given by

$$\int_0^\infty \frac{1}{B_1}\exp\left(-\frac{1}{1-\rho^2}B_1-\frac{\eta}{1-\rho^2}\frac{1}{B_1}\right)dB_1 = \int_0^\infty \frac{1}{x}\exp\left(-\frac{x}{2}-\frac{4\eta}{(1-\rho^2)^2}\frac{1}{2x}\right)dx \tag{6.73}$$

In term of the integral formula [39, p. 712, 6.653, Eq. 1] shown as

$$\int_0^\infty \frac{1}{x}\exp\left(-\frac{1}{2}x-\frac{1}{2x}(a^2+b^2)\right)I_v\left(\frac{ab}{x}\right)dx = \begin{cases} 2I_v(a)K_v(b) & 0<a<b, \mathrm{Re}v>-1 \\ 2K_v(a)I_v(b) & 0<b<a, \mathrm{Re}v>-1 \end{cases} \tag{6.74}$$

we obtain exactly (6.51) by combining (6.72), (6.73) and (6.74) with $I_0(0) = 1$ and the assumption that $v = 0$, $a = \frac{2\sqrt{\eta}}{1-\rho^2}$, $b = 0$.

Appendix 6.3: The Approximate PDF for ξ

Via the following asymptotic formulas of the modified Bessel functions [40, p. 511, Eqs. 9.56b and 9.56d],

$$\begin{aligned} I_v(x) &= \frac{e^x}{\sqrt{2\pi x}}\left[1 + O\left(\frac{1}{x}\right)\right] \approx \frac{e^x}{\sqrt{2\pi x}} m_1 \\ K_v(x) &= \sqrt{\frac{\pi}{2x}} e^{-x}\left[1 + O\left(\frac{1}{x}\right)\right] \approx \sqrt{\frac{\pi}{2x}} e^{-x} m_2 \end{aligned} \tag{6.75}$$

where m_1 and m_2 are both a constant closed to 1. The expression $O\left(\frac{1}{x}\right)$ means an infinitesimal quantity of the same order as $\frac{1}{x}$. Herein, Utilizing the approximation shown in (6.75), and thus (6.52) becomes

$$\begin{aligned} p_\varsigma(\varsigma) &= \frac{4\varsigma^n n^{n+1}}{\Gamma(n)(1-\rho^2)\rho^{n-1}} \sqrt{\frac{1-\rho^2}{4\pi n\rho\varsigma}} \exp\left(\frac{2n\rho}{1-\rho^2}\varsigma\right) m_1 \sqrt{\frac{\pi(1-\rho^2)}{4n\varsigma}} \exp\left(-\frac{2n}{1-\rho^2}\varsigma\right) m_2 \\ &= \frac{m_1 m_2}{\rho^{n-1/2}} \frac{n^n}{\Gamma(n)} \varsigma^{n-1} \exp\left(-\frac{2n\varsigma}{1+\rho}\right) \end{aligned} \tag{6.76}$$

Owing to $\int_0^\infty p_\varsigma(\varsigma)\mathrm{d}\varsigma = 1$, $\frac{m_1 m_2}{\rho^{n-1/2}} = \left(\frac{2}{1+\rho}\right)^n$. Letting $\beta = \frac{2n}{(1+\rho)\sigma_1\sigma_2}$, one can arrive at (6.53) from (6.76).

References

1. A. Lopès, F. Séry, Optimal speckle reduction for the product model in multilook polarimetric SAR imagery and the Wishart distribution. IEEE Trans. Geosci. Remote Sens. **35**(3), 632–647 (1997)
2. J.S. Lee, M.R. Grunes, G.D. Grandi, Polarimetric SAR speckle filtering and its implication for classification. IEEE Trans. Geosci. Remote Sens. **37**(5), 2363–2373 (1999)
3. J.M. Beaulieu, R. Touzi, Segmentation of textured polarimetric SAR scenes by likelihood approximation. IEEE Trans. Geosci. Remote Sens. **42**(10), 2063–2072 (2004)
4. J.S. Lee, M.R. Grunes, T.L. Ainsworth, L.J. Du, D.L. Schuler, S.R. Cloude, Unsupervised classification using polarimetric decomposition and the complex Wishart classifier. IEEE Trans. Geosci. Remote Sens. **37**(5), 2249–2258 (1999)
5. J. Schou, H. Skriver, A.A. Nielsen, K. Conradsen, CFAR edge detector for polarimetric SAR images. IEEE Trans. Geosci. Remote Sens. **41**(1), 20–32 (2003)
6. N.R. Goodman, Statistical analysis based on a certain multivariate complex gaussian distribution (an introduction). Ann. Math. Stat. **34**(152), 152–180 (1963)
7. S.H. Yueh, J.A. Kong, J.K. Jao, R.T. Shin, L.M. Novak, K-distribution and polarimetric terrain radar clutter. J. Electromagn. Waves Appl. **3**(8), 747–768 (1989)
8. L.M. Novak, M.B. Sechtin, M.J. Cardullo, Studies of target detection algorithms that use polarimetric radar data. IEEE Trans. Aerosp. Electron. Syst. **AES-25**, 150–165 (1989)
9. J.S. Lee, K.W. Hoppel, S.A. Mango, A.R. Miller, Intensity and phase statistics of multilook polarimetric and interferometric SAR imagery. IEEE Trans. Geosci. Remote Sens. **32**(5), 1017–1028 (1994)

10. J.S. Lee, D.L. Schuler, R.H. Lang, K.J. Ranson, K-distribution for multi-look processed polarimetric SAR imagery, in *Proceedings of the IEEE IGARSS*, Pasadena, CA, vol. 4 (1994), pp. 2179–2181
11. C.C. Freitas, A.C. Frery, A.H. Correia, The polarimetric G distribution for SAR data analysis. Environmetrics **16**(1), 13–31 (2005)
12. L. Bombrun, J.M. Beaulieu, Fisher distribution for texture modeling of polarimetric SAR data. IEEE Geosci. Remote Sens. Lett. **5**(3), 512–516 (2008)
13. A.C. Frery, H.J. Muller, C.C.F. Yanasse, S.J.S. Sant'Anna, A model for extremely heterogeneous clutter. IEEE Trans. Geosci. Remote Sens. **35**(3), 648–659 (1997)
14. L. Bombrun, G. Vasile, M. Gay, F. Totir, Hierarchical segmentation of polarimetric SAR images using heterogeneous clutter models. IEEE Trans. Geosci. Remote Sens. **49**(2), 726–737 (2011)
15. H.C. Li, W. Hong, Y.R. Wu, P.Z. Fan, On the empirical-statistical modeling of SAR images with generalized gamma distribution. IEEE J. Sel. Top. Signal Process. **5**(3), 386–397 (2011)
16. H.C. Li, W. Hong, Y.R. Wu, Generalized gamma distribution with MoLC estimation for statistical modeling of SAR images, in *Proceedings of the APSAR*, (2007) pp. 525–528
17. J. M. Nicolas, Stian Normann Anfinsen (translator), Introduction to second kind statistic: application of log-moments and log-cumulants to SAR image law analysis. Trait. Signal **19**(3), 139–167 (2002)
18. C. Tison, J.M. Nicolas, F. Tupin, H. Maitre, A new statistical model for Markovian classification of urban areas in high-resolution SAR images. IEEE Trans. Geosci. Remote Sens. **42**(10), 2046–2057 (2004)
19. V.A. Krylov, G. Moser, S.B. Serpico, J. Zerubia, On the method of logarithmic cumulants for parametric probability density function estimation. INRIA, Res. Rep. **7666**, 1–36 (2011)
20. G. Moser, J. Zerubia, S.B. Serpico, SAR amplitude probability density function estimation based on a generalized Gaussian model. IEEE Trans. Image Process. **15**(6), 1429–1442 (2006)
21. H.C. Li, W. Hong, Y.R. Wu, P.Z. Fan, An efficient and flexible statistical model based on generalized gamma distribution for amplitude SAR images. IEEE Trans. Geosci. Remote Sens. **48**(6), 2711–2722 (2010)
22. S.N. Anfinsen, Statistical analysis of multilook polarimetric radar images with the Mellin transform, Ph.D. dissertation, Univ. Tromsø, Tromsø, Norway, 2010
23. S.N. Anfinsen, T. Eltoft, Application of the matrix-variate Mellin transform to analysis of polarimetric radar images. IEEE Trans. Geosci. Remote Sens. **49**(6), 2281–2295 (2011)
24. A.C. Frery, A.H. Correia, C.C. Freitas, Classifying multifrequency fully polarimetric imagery with multiple sources of statistical evidence and contextual information. IEEE Trans. Geosci. Remote Sens. **45**(10), 3098–3109 (2007)
25. A.M. Mathai, Distribution of the canonical correlation matrix. Ann. Inst. Stat. Math. **33**(pt. A), 35–43 (1981)
26. C.H. Gierull, I.C. Sikaneta, Estimating the effective number of looks in interferometric SAR data. IEEE Trans. Geosci. Remote Sens. **47**(11), 1733–1742 (2002)
27. S.N. Anfinsen, A.P. Doulgeris, T. Eltoft, Estimation of the equivalent number of looks in polarimetric synthetic aperture radar imagery. IEEE Trans. Geosci. Remote Sens. **47**(11), 3795–3809 (2009)
28. N. Batir, On some properties of digamma and polygamma functions. J. Math. Anal. Appl. **328**(1), 452–465 (2007)
29. D.J. Crisp, The state-of-the-art in ship detection in synthetic aperture radar imagery. DSTO, Dept. Defence, Australian Government, Canberra, Australia, Public Release Document DSTO-RR-0272, 2004
30. C.J. Oliver, S. Quegan, *Understanding Synthetic Aperture Radar Images* (Artech House, Norwood, MA, 1998)
31. I.C. Sikaneta, Detection of ground moving objects with synthetic aperture radar. Ph. D. thesis. University of Ottawa, 2002
32. F.T. Ulaby, C. Elachi, *Radar Polarimetric for Geoscience Application* (Artech House, Boston, 1990)

33. C. Livingstone, A. Thompson, The moving object detection experiment on RADARSAT-2. Can. J. Remote. Sens. **30**(3), 355–368 (2004)
34. V.A. Krylov, G. Moser, S.B. Serpico, J. Zerubia, Supervised high resolution dual polarization SAR image classification by finite mixtures and copulas. IEEE J. Sel. Top. Sign. Proces. (2012, In Press)
35. R.D. Chaney, M.C. Bud, L.M. Novak, On the performance of polarimetric target detection algorithms. IEEE AES Magazine 10–15 (1990)
36. Gierull, Statistics of SAR interferograms with application to moving target detection. *DREO Technical Report TR 2001-045*, Defense Research Establishment Ottawa, Department of National Defense, Ottawa, Canada, 2001
37. A.C. Frery, J. Jacobo-Berlles, J. Gambini, M. Mejail, Polarimetric SAR image segmentation with B-Splines and a new statistical model. Multidimension. Syst. Signal Process. **21**, 319–342 (2010)
38. J.M. Nicolas, Introduction to second kind statistic: application of log-moments and log-cumulants to SAR image law analysis. Trait. Signal **19**(3), 139–167 (2002)
39. S.I. Gradshteyn, I.M. Ryzhik, *Table of Integrals, Series, and Products*, 7th edn. (Academic Press, San Diego, CA, 2007)
40. I.N. Bronshtein, K.A. Semendyayev, G. Musiol, H. Muehlig, *Handbook of Mathematics*, 5th edn. (Springer, 2007)

Printed by Printforce, the Netherlands